STAR MAGIC

ABOUT THE AUTHOR

Sandra Kynes is a yoga instructor, Reiki practitioner, and a member of the Bards, Ovates and Druids. She likes developing creative ways to explore the world and integrating them with her spiritual path, which serves as the basis for her books. She has lived in New York City, Europe, England, and now coastal New England. She loves connecting with nature through gardening, hiking, and ocean kayaking. Visit her website at www.kynes.net.

Sandra Kynes

STAR MAGIC

The Wisdom of the Constellations for
Pagans & Wiccans

Llewellyn Publications
Woodbury, Minnesota

First Edition
First Printing, 2015

Cover design by Lisa Novak
Cover Illustration: Erika Steiskal
Interior illustrations by Gavin Dayton Duffy and the Llewellyn Art Department

Llewellyn Publications is a registered trademark of Llewellyn Worldwide Ltd.

Library of Congress Cataloging-in-Publication Data

Kynes, Sandra, 1950–
  Star magic : the wisdom of the constellations for pagans & wiccans / by Sandra Kynes. — First Edition.
     pages   cm
  Includes bibliographical references and index.
  ISBN 978-0-7387-4169-7
  1. Stars—Miscellanea. 2. Constellations—Miscellanea. 3. Astrology. I. Title.
  BF1724.3.K96 2015
  133.5—dc23
                                                            2014017317

Llewellyn Publications
A Division of Llewellyn Worldwide Ltd.
2143 Wooddale Drive
Woodbury, MN 55125-2989
www.llewellyn.com

Printed in the United States of America

Other Books by Sandra Kynes

A Year of Ritual

Change at Hand

Gemstone Feng Shui

Llewellyn's Complete Book of Correspondences

Mixing Essential Oils for Magic

Sea Magic

Whispers from the Woods

Your Altar

This book is dedicated to stargazer Lyle Koehnlein,
my son and special star.

CONTENTS

INTRODUCTION

Have you ever wished upon a star, reached for the stars, or thanked your lucky stars? Those twinkling lights in the dark heavens have mystified and bedazzled people since time immemorial, and they still fascinate us today. Although the stars may seem remote, their familiar patterns have provided comfort and guidance when the world was lit only by firelight. Many ancient people observed the night sky and documented the constellations in one form or another; some through myth and others using complex mathematics. People of the past used the stars to reckon time, navigate the seas, and coordinate the planting and harvesting of crops. Even today, in our well-lit techno world, the stars remain a source of wonder, drawing enormous attention whenever a spacecraft sends images home to Earth.

I have fond childhood memories of summer nights lying in the grass, looking at the sky, and trying to fathom the distances. My father was a Boy Scout Scoutmaster and loved spending time outside teaching young people about the wonders of nature. Of course, that included my sister and me. On nights of meteor showers, he would point out the areas of the sky to watch. I would secretly cross my fingers and hope to see a shooting star upon which to wish. The enthusiasm that I inherited from my father was passed on to my son. During his childhood, he and I spent many Saturdays uptown at the American Museum of Natural History in New York City. The dinosaur room always topped our agenda, but the day wouldn't be complete without visiting the Hayden Planetarium where we both marveled at the spectacle even though it was only a show. My son's interest continues to this day, and as a professional photographer he takes every opportunity to capture the stars.

The most well-known and impressive testament to ancient peoples' observations of the sky is Stonehenge on the Salisbury Plain in England. Not only does it speak of the importance in marking celestial events, but also the cumulative knowledge necessary to embark on such a project. However, while the ancients were quite familiar with the patterns in the night sky, we, sadly, are not. We are drawn by the glamour of the moon as we celebrate its shining fullness with our esbats. And although we may follow the turning wheel of the zodiac, how many of us can identify these constellations when we look up at the night sky?

Beyond recognizing the Big or Little Dippers and maybe Orion's Belt, what star patterns do most of us know? Perhaps part of the problem is that few of us experience the true darkness of night as our ancestors did. Their world wasn't as bright as ours with its artificial light 24-7. Light itself is a form of pollution that takes away the dark cover of night and diminishes or obscures the twinkling grandeur above. As a result, the most inspiring sight to behold in the night sky is the moon. While Earth's satellite is enchanting, we unfortunately miss almost everything else, or when we do see the stars we may not know what we are looking at. Because of this, we are missing out on some very powerful magic.

The idea of using constellations for magic and ritual isn't new. Medieval texts included details about stars and how to determine the optimal time to draw their influence into talismans as well as other uses. A small remnant of this remains in our use of birthstone jewelry. If we believe that the constellations of the zodiac hold a great deal of power and influence, then why not the others? While each of the zodiac signs have their time on the solar stage of day, have you wondered what goes on opposite them in the night sky? When Pisces and Aries are casting their birth influences as backdrops to the sun, Libra and Virgo are bringing balance and welcoming spring during the night. In addition, just as a cloudy night does not block the energy of a full moon, so too does the power of stars reach us even if we cannot see them.

The stars have always had a profound influence on us. Many of the goddesses we acknowledge, worship, and honor today have been known as star goddesses. Known by many names in numerous cultures, Astarte was ultimately known as the Queen of Heaven.

Ishtar's symbol was the eight-pointed star, and Inanna was known as the Queen of Heaven and Earth. Legends of the Egyptian goddess Nut, who is depicted as the span of heaven arching over her earth husband Geb, predate sacred texts. Isis, first-born daughter of Nut and Geb, became the mistress of the cosmos when her parents "retired to heaven." [1] The Romans depicted Vesta as a star goddess whose pure flame was a beacon in the darkness of the night sky.

According to Sumerian texts, the stars were home to the gods of creation. The Egyptians equated terrestrial geography with celestial fields and cities. To the people of India, earthly cities had heavenly counterparts, too. All royal cities in India were based on a mythical celestial city. In many cultures, temples were regarded as sacred mountains and the meeting point of heaven and earth. Mesopotamian ziggurat temples represented cosmic mountains upon which the heavenly gods could descend to Earth.

Although there have been many permutations over the years, the widely used *Charge of the Goddess* refers to the Star Goddess. The original version has been attributed to Charles Leland, Gerald Gardner, and Doreen Valiente, and it contains: "Hear ye the words of the Star Goddess, she in the dust of whose feet are the hosts of heaven, and whose body encircles the universe." [2]

Another reason that Wiccans and Pagans may want to take more of an interest in the constellations is the simple fact that our basic and most-recognizable symbol, the pentagram, is a star. We follow an earth-centered spirituality, yet we look to the heavens. Being outside under a limitless canopy of stars invites us to open our souls and connect with something far, far greater than ourselves. Just as we can draw down the energy of the moon, so too can we tap into the celestial energy beyond.

I am very much the amateur astronomer. I have not had formal training, but I love jumping in and learning as much as I can about the things that interest me. In addition to my father's enthusiasm, growing up at the beginning of the space age sparked the stargazer in me. While I never wanted to go into space, I wanted to know what was beyond this planet. There's something comforting about looking at the night sky and having a pretty good idea of what is where no matter where I am. It's like being in a familiar neighborhood. In my teen years, as I found my way onto a Pagan path, my perspective on the natural world evolved into a reverence that was highlighted by awe. To me, the term "natural world" wasn't limited to the earth; it always included the dome of the sky. Celebrating the esbats is about more than just the moon for me; it is also about that mysterious, endlessness beyond. Despite the enormity of the universe, I have a sense of place because

I can find the markers of the seasons. This sense of place also provides a connection with my ancestors because they would have seen what I see. This stellar continuity is a spiral that connects us. The stars also connect us with the magic of the universe. Truly as above, so below. Over the years I accumulated quite a few notes on my thoughts, ideas, and experiences, which I have decided to share.

This book explores the night sky, examines the mythology of the constellations, and presents a new interpretation that is relevant for twenty-first-century Pagans and Wiccans. The stars connect us with the past, and chapter 1 begins with a historical view on how people of ancient times regarded the constellations. In addition, we will see how modern astrology is based on the zodiac as it was observed several thousand years ago and we will learn how these constellations are different today. After introducing celestial coordinates, chapter 2 will get you started reading star maps. It also includes information on planispheres and a few online resources and smartphone apps. We will also learn about official star designations and how "a" star is sometimes more than one.

This book is also about magic and magic is about moving and using energy. If you are familiar with drawing down the moon for ritual or magic, you are well on your way to working with the stars. Chapter 3 provides an introduction to general energy work and methods for drawing on and using energy from the stars to enhance rituals and boost magic.

Because the Wheel of the Year turns the earth and heavens, this book is designed to allow you to start with the chapter that coincides with the current season. One thing to keep in mind, however, is that the visible constellations do not suddenly change. In early spring, we will still see some of the late winter constellations and some of the late spring constellations will not be visible. Like the changing seasons on Earth, it is a gradual process. Chapters 4 through 7 cover the seasonal constellations visible from the Northern Hemisphere.

While chapter 8 is entitled *The Southern Hemisphere*, readers in the Northern Hemisphere should not immediately assume it is not for them. In fact, a number of constellations included in chapters 4 through 7 are actually southern. The additional southern constellations presented in chapter 8 are also visible in many parts of the north. Because what we are able to see is based on where we are located on Earth, I have included the latitudes between which the constellations can be seen. For reference, appendix A provides latitude information for a number of cities around the world.

These chapters include star maps to help familiarize you with the night sky according to season. I am not a cartographer, so these maps are approximations that just show the basic position of the constellations and how they relate to each other. Each individual entry includes a depiction of some of the stars in that constellation and where they fit within the imagined figure. Rather than reproduce entire star patterns, these individual maps are intended to suggest the constellations. Keeping these simple makes it easier to re-create a star pattern for ritual and magic work. The many star charts that you may find in other books and online often provide different renderings of constellations, and you may be inspired to create your own. For this reason, I have included the official designation of each star in the drawings to aid you in orienting it to a more complete map of its constellation. The drawings also serve as a map for coordinating each star's color, should you choose to combine color magic with your star work.

Information on each constellation includes its history, associated myths, and the particulars on notable stars. Each entry includes the constellation's official name and its common English name. Also included is information on the pronunciation of constellation and star names and the constellation abbreviations that you will find on star maps. Of course, each constellation entry includes an interpretation for Pagan and Wiccan magic and ritual. These sections provide ideas and details on how to apply the energy of a constellation to your life. My hope is that this will spark your creativity and you will find ways to make star magic uniquely yours.

As mentioned, this book works with the constellations and actual seasons rather than the time frames used in modern astrology. Despite this difference, star magic and astrology are not incompatible. While star magic is based on the current position and movement of constellations, astrology uses a historical system that holds a great deal of meaning. Although I am not an astrologer, I believe that practitioners can draw energy from the constellations that appear in the night sky to enhance and support their work.

For astrologers, I have included information on the planets associated with particular constellations and individual stars. For nonastrologers, appendix C provides a table of planet energies and qualities. In addition, appendix C contains details about the "fixed stars" that were considered powerful guides during the Middle Ages and Renaissance.

This book can be used in a number of ways. It can bring awareness of the night sky beyond the powers of the moon to enhance your esbat rituals. It can serve as an introduction to stargazing with a magical twist, and it may even propel you further into studying astronomy. More simply, it will bring you that neighborhood feeling when you step outside at night and see familiar star patterns. Most of all, you will learn which stars are overhead throughout the seasons and how to use their energy in your life.

No matter how you want to engage in star magic, spending a little time outside at night provides a different and refreshing perspective. Night softens the world—bringing rest as well as incubation for magic and creativity. Stargazing requires us to stop and look. This simple act allows us to reach inside our souls and experience the wonder that echoes down through the eons from people in the far distant past. The energy of the stars envelops our planet and holds us in the web of the cosmos. In the scheme of the universe we are so tiny, yet we are a part of something so vast and wondrous.

Like gazing at clouds, we may each perceive something different in the constellations. This book is my interpretation, but I encourage you to follow your feelings and intuition. Trust what you see in the stars, and let them guide you to a new level of magic and wonder.

Chapter One

Astronomy and Astrology

Historical Background

Unlike people of ancient times, we tend to stay in our own backyard of the solar system. This strikes me as odd for a civilization with space telescopes that allow us to see farther than our ancestors ever dreamed. In this book, we will explore the constellations, learn how to look for them, and discover new ways to bring star magic into our lives.

Modern astronomy recognizes eighty-eight constellations, some of which are based on the forty-eight ancient Greek star figures described by the mathematician Claudius Ptolemaeus (circa 100–170 CE), better known simply as Ptolemy. The other forty modern constellations are based on European star atlases from the fifteenth through seventeenth centuries. Although the Latin names of many constellations were derived from their earlier Greek designations, not all of them originated with Greek astronomers. While Ptolemy cataloged more than 1,000 stars, he associated them with constellations that had been known for centuries, many of which originated with Sumerian and Babylonian astronomers.

Ptolemy's knowledge of astronomy is believed to have come indirectly from the Egyptians, Chaldeans, and Phoenicians. His information was more directly derived from the work of Greek mathematician Eudoxus of Cnidus (circa 390–340 BCE) to whom scholars have attributed the earliest text on the constellations. Eudoxus traveled widely to study a number of disciplines, including medicine and law. While his work itself does not survive, his star figures were known through the writings of Greek poet Aratus (circa 315–245 BCE). Aratus's literary work entitled *Phaenomena* was often used as an introduction to astronomy and is the oldest description of the constellations that has come down through the centuries intact.

Other early stargazers and chroniclers were the Greek mathematician Eratosthenes (circa 276–194 BCE), who described forty-two constellations, and the astronomer Hipparchus (circa 170–127 BCE), who noted forty-eight groups of stars. It was Hipparchus who discovered the ongoing shifting of the stars that we know today as "precession," or "precession of the equinoxes." Regardless of source, the Greeks wrapped the star figures in their own mythology and created an ongoing drama in the night skies. Arab, Persian, Roman, and later medieval European astronomers also adopted this interpretation of the constellations.

While Greek influence on astronomy was widespread, the Greeks were not the earliest stargazers. The Babylonians, Persians, Egyptians, Indians, and Chinese also studied the heavens. As early as 2300 BCE the Chinese were coordinating their twelve zodiac signs with twenty-eight lunar mansions.[3] The Hindus of India had a similar system by which they tracked the moon's progress across the stars, and they associated a particular star with each lunar mansion. The Assyrians regarded the stars as divinities with either benevolent or malevolent powers.

In later times, as Europeans made contact with tribal people around the world, it was discovered that they also observed the stars and made it part of their lore. The Bushmen of Africa considered the stars to have been people or animals that once walked the earth. In New Zealand, the stars were regarded as legendary heroes whose brightness depended on their greatness in terrestrial battle before passing on to the heavens. As well as being regarded as the children of the sun and moon by the early people of Peru, the stars were believed to function as guardian deities. And last but not least, many Native American tribes looked on the constellations as celestial counterparts to powerful terrestrial animals.

Most early cultures observed the seasonal differences in the night sky as the star patterns changed over the course of a year. Providing a longer count than the moon, reckoning by the stars enabled people to mark time for planting, harvesting, and ritual. Visible year round, the circumpolar stars and especially the North Star—that reliable beacon that never moved—provided markers for navigation through the night seas. Building on what had gone before, the Greeks also used the constellations for navigation and agriculture. This is evident in the works of the writer Homer (eighth century BCE) who described

Odysseus, the hero of his stories, navigating by the stars. Additionally, the poet and farmer Hesiod (circa 700 BCE) advised others on how to use the stars for determining the right time to plant their fields. The aesthetic beauty of the constellations also served as a decorative motif for the Greeks, who painted them on domed ceilings, and the Romans, who worked them into tapestries with threads of gold and silver.

Based on a set of tablets dating to approximately 1800 BCE, scholars believe that the Babylonians knew how to use calculations to fill in for the times when observations could not be made due to weather conditions or other circumstances. These tablets span one thousand years and are thought to be a type of calendar that includes listings of stars along with observation dates and times.[4] Using this information and incorporating the divination results of their stargazing into their daily lives developed into a system that is considered to have been far more complex than our present-day astrology. Using the stars for in-depth horoscopes to guide daily life was not the only important aspect for some people in ancient Mesopotamia. During the first and second centuries, Sabianism, or star worship, sprang up in the city of Harran. This early Sabianism has been described as involving astrology, star worship, and magic. Although Sabianism was part of the pre-Islamic religion of the Harranians, the term *Sabian* was later applied to any form of star worship.

The Egyptian's system of celestial observation dates to the Old Kingdom (circa 2650–2150 BCE); however, it was not as complex as the one used by their northern neighbors in Babylon. Even so, their star calendar contained thirty-six groups of stars. Within eight centuries from the initiation of their sky studies, the Egyptians were building temples and monuments that aligned with certain stars. In addition, they divided the night sky in half with Meskhet, the Big Dipper, as the northern marker and Sopdet, the star Sirius, as the southern marker. The constellation Orion was known as the Guardian of the Soul of Horus.

While the classical civilizations left records of their stargazing, little was known about any such endeavors by northern Europeans, where there were no great cities and no early forms of writing. Germany was generally considered the dark heart of Europe, primitive and uncivilized. However, the discovery of the Nebra sky disc in 2001 has changed that presumption. Found in central Germany, this bronze disc has a diameter of twelve and a half inches and it is embossed with gold leaf depictions of the crescent and full moons, the sun, and stars. It has been dated to 1600 BCE. After considerable study, scholars believe that Bronze Age Europeans were far more sophisticated than previously thought. According to astronomer Wolfhard Schlosser of the Ruhr University at Bochum, Germany, Bronze Age sky gazers knew as much as the Babylonians. In addition, their astronomical knowledge and abilities allowed them to use a combination of solar and lunar calendars.

With some surprise, scientists realized that the small group of seven stars on the Nebra sky disc actually depicts the Pleiades, a cluster of stars in the constellation Taurus. Realistic star images did not appear until 1400 BCE in Egypt, making the Nebra sky disc the oldest accurate picture of the night sky. In addition, the mysterious golden horizon bands that were originally thought to represent some sort of solar boat turn out to be markers for the solstices. Actually, this is not so startling because all across northern Europe many standing stones erected in prehistoric times were aligned to mark the solstices. According to Professor Miranda Green of the University of Wales, the Nebra sky disc presents a mosaic of symbols that were part of a complex European-wide belief system. I found it wonderful to learn that the early Pagans of Europe who erected stone circles and alignments to celebrate the solstices were also gazing beyond to the stars.

After the decline of the Roman Empire, Europe entered the chaotic period known as the Dark Ages (approximately 476–800 CE), when the advancement of learning and knowledge came to a virtual halt. Luckily the observations and ideas of many ancient stargazers were translated and kept alive by scholars in the Middle East. A great deal of this work was re-translated from Arabic to Latin during the twelfth and thirteenth centuries as it was carried back into Europe. However, during the Dark Ages, Ireland was a little bright spot in Europe that quietly maintained a high level of culture and learning.

Writing about early Irish astrology, distinguished Celtic scholar, historian, and author Peter Berresford Ellis noted that the Irish, and Celts in general, had a long tradition of astrological study. As the first-century-BCE Coligny Calendar demonstrates, Celtic cosmology was highly sophisticated. In addition, the ancient Brehon Laws of Ireland required that astronomers/astrologers prove their level of qualification in order to practice their art. Prior to Arabic texts on astronomy and astrology being translated into Irish in the twelfth century, the earliest writings in Ireland reveal a native concept when naming the constellations and planets. For example, the constellation Leo was called *An Corran*, "the reaping hook," which describes the sickle shape of stars that define Leo's head.[5] Ellis also noted that the astronomical information on stars, comets, and other celestial bodies recorded in the various annals and chronicles of Ireland were more accurate than many others produced in Europe at that time.

With all the knowledge that had been accumulated in the ancient world, it was Ptolemy's work that stood the test of time until the fifteenth century when scientific study took off. Polish mathematician and astronomer Nicolaus Copernicus (1473–1543) put forth his theory that the sun was the center of the universe and not Earth. Although his idea was controversial at the time, it marked the beginning of a change in the way people viewed Earth and its place among the stars. Danish nobleman and astronomy enthusiast Tycho Brahe (1546–1601) built an observatory and, without a telescope, accurately mapped almost eight hundred stars. In 1603, German astronomer Johann Bayer (1572–1625) published a star atlas that was the first to map the entire sky. Bayer also established the naming convention for stars within a constellation that is still used today. Another important astronomer of the time was Johannes Kepler (1571–1630), who put forth the laws of planetary motion.

While modern scientific study of the stars was getting started during this period, there was no distinction between astronomy and astrology as there is today. In fact, the great founders of modern astronomy were also astrologers. It is somewhat ironic that during the scientific revolution astrology remained an integral part of astronomy, mathematics, and medicine. In general practice, medieval astrology was divided between the two applications of divination and medicine. The famed English herbalist Nicholas Culpeper (1616–1654) also wrote several books on astrology and integrated his astrological knowledge with his herbal practice. From a medical standpoint, various parts of the body were believed to be under the influence of the constellations of the zodiac. Because of this, a physician did not diagnose a problem until he performed a series of complex astrological computations. The end result would determine whether or not particular cosmic influences would be advantageous for the patient. Eventually the scientific thinking that developed after the seventeenth century demanded observable physical explanations. While this moved astrology outside the realm of science, its influence and popularity has not waned.

As mentioned, the work of ancient stargazers was kept alive in the Middle East, and many of the traditional names of stars that we know today are a mix of Arabic, Greek, and Latin as well as mistranslations from one language to another. The star Rigel in the Orion constellation is one such case. Instead of the hunter Orion, Arab astronomers saw this constellation as a female figure they called al-Jauza and named the bright star that marked her left foot Rijl al-Jauza, "the foot of al-Jauza." Over the centuries and through several translations, Rijl evolved into Rigel. The spelling varied as well, which is why we find the same word spelled *rigil* as in the star name Rigil Kentaurus, "the foot of the centaur." Also, many of the star names beginning with "al" stem from the Arabic "al," which is the equivalent of the word "the" in English.

As we study the stars, we find that there are no older constellations in the far Southern Hemisphere. This is because that area of the sky was unknown to observers in the civilizations that discerned and named these star patterns. The classical constellations are centered on the north celestial pole as it was over 2,000 years ago. The north celestial pole is above Earth's terrestrial North Pole, and, likewise, the south celestial pole is above the South Pole. Earth's axis is tilted; it is not straight up and down. As a result, the terrestrial and celestial poles are at corresponding angles that shift over time due to Earth's rotation and wobble. As our planet spins, it also wobbles like a top. The name for this wobble is "precession," and it is caused by the gravitational pull of the moon and sun, which shifts Earth's orientation in space.

If an imaginary line from the axis were extended outward into space, it would align with different stars over time. Because of precession, Polaris is now the nearest star to the North Pole and it marks the north celestial pole. Also known as the North Star and the Pole Star, Polaris is calculated to move closer to and within one degree of the North Pole around the year 2100. Also due to precession, approximately 4,500 years ago the star Thuban in Draco the Dragon constellation had that honor. While the Southern Hemisphere does not have a pole star, the nearest star to the south celestial pole is Beta Hydri in Hydrus the Southern Water Snake.

Precession is also known as "precession of the equinoxes" because Earth's shift in space affects which constellation marks the spring equinox. This slow shift makes the sun appear to be lagging behind in its progression through the zodiac and over time the constellation that marks the spring equinox lags behind. The change in constellation at the spring equinox gave rise to the concept of "ages" or "astrological ages," so when there is a constellation change at the spring equinox it is the beginning of that age. The Age of Aquarius is an example that most people have heard about. The full cycle of precession takes almost 26,000 years. Today the spring equinox occurs when Pisces marks the equinox, however; 2,500 years ago it was Aries. Earlier still, Taurus would have been the constellation at the equinox, which is thought to be one of the reasons that bulls have been associated with spring fertility rites.

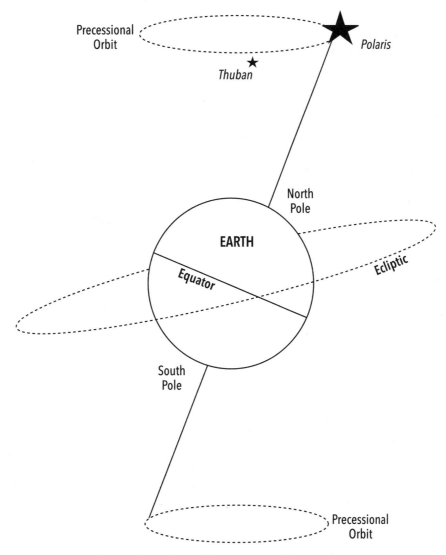

Figure 1.1. Earth's wobble causes the shift in pole stars.

In 1922, the International Astronomical Union (IAU) established the modern eighty-eight constellations and introduced the convention whereby defined regions were created to cover the entire sky without overlaps or gaps. To accomplish this, newer constellations were added to fill in the spaces between some of the classical ones. Because of this reorganization, some constellations such as Argo Navis became extinct. However, this is not to say that the stars went the way of the dinosaurs; instead, certain star groups were no longer recognized as official constellations.

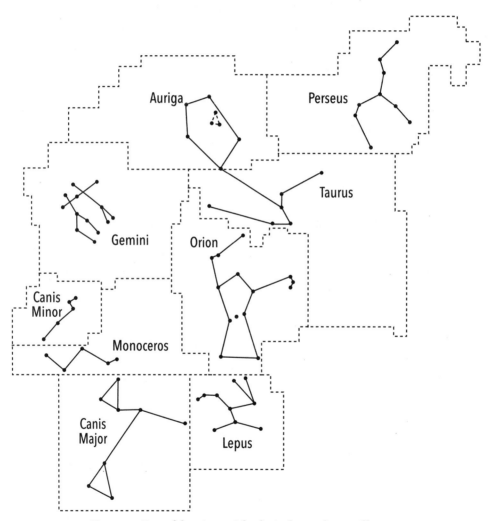

Figure 1.2. Part of the winter night sky is shown above to illustrate how defined regions cover the entire sky without overlaps or gaps.

The word *constellation* means "a gathering of stars"; however, in today's terms a constellation is an acknowledged pattern of stars within a defined region of the sky. Although the stars within a constellation form a pattern, they often have no relation to each other and, in fact, are at different distances from Earth. These stars may appear to be of equal distance because of their varying brightness. In other words, a very bright star that is farther away from Earth may seem to be the same distance as a closer one that is less brilliant.

While an asterism is also a discernible pattern of stars, it is not officially recognized as a constellation. However, asterisms are often helpful for finding other stars and constellations. The Big Dipper is a well-known example of an asterism in the constellation Ursa Major, the Great Bear.

Of the eighty-eight constellations, only a few are members of the zodiac, which are also known as sun signs. When the sun's position is between Earth and these constellations, their stars become a background along the narrow path of the sun's annual arc across the sky. The scientific name for this path is the ecliptic. Of course, it is Earth that is moving around the sun, but to us the sun appears to be moving. To say that the sun is in Pisces means that this constellation is in the same direction as the sun and would be like a background for the sun if we were able to see Pisces during the day. Figure 1.3 shows that the daytime view from Earth on the vernal equinox puts Pisces on the other side of the sun.

Whatever constellation is the current backdrop to the sun, its opposite on the ecliptic will be visible in the night sky. This is because as Earth spins it faces away from the sun at night and thus away from the current zodiac sign. For example, during August when Leo is the sun's background constellation, Aquarius can be seen at night. In late February, these positions are reversed and Leo is seen at night. Also appearing near the ecliptic are the moon and planets, which change location in the sky more quickly than do stars.

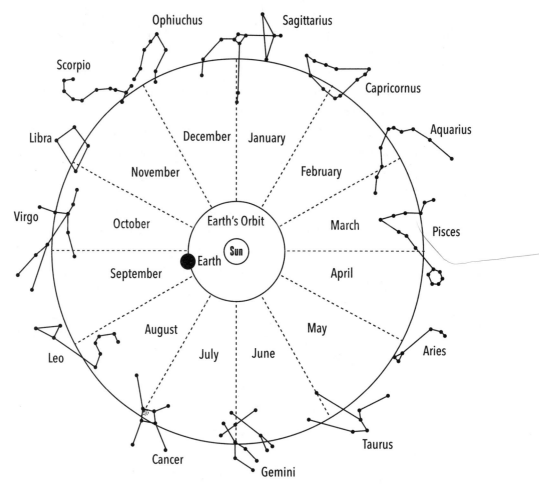

Figure 1.3. The constellations of the zodiac are along the sun's path called the ecliptic.
On the vernal equinox in March, Pisces is on the opposite side of the sun from Earth.

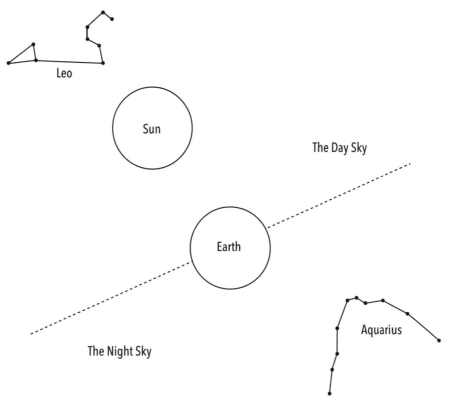

Figure 1.4. The constellations of the day and night sky are different.
They are opposite each other on the circle of the ecliptic.

Modern astrology uses the twelve constellations and dates that were established more than two thousand years ago. As a result, Aries is the first sign of the zodiac in astrology, and its time begins at the spring equinox. However, as previously mentioned, Pisces is the actual constellation in position on the ecliptic today. While the zodiac used in astrology is composed of fairly equal segments of approximately thirty days, the actual time that each constellation is the sun's backdrop varies because some constellations are larger than others. Also as previously mentioned, a constellation is a region of sky and not just the star pattern itself. In some instances, the ecliptic passes through just a small part of a constellation/region of sky, which also affects the amount of time it serves as a backdrop to the sun. Virgo has the most time with forty-five days, and Scorpius the least at just seven days.

In fact, there are now thirteen constellations that fall within the ecliptic. Table 1.1 lists the constellations on the ecliptic, their actual dates, and the number of days they serve as a backdrop to the sun. Also included are the traditional astrological dates, which shift slightly, give or take one day each year. The actual dates and days as a backdrop listed in this table come from Dr. Lee T. Shapiro, the Director of the Morehead Planetarium at the University of North Carolina in Chapel Hill. These were made available on one of the National Aeronautics and Space Administration's (NASA) websites (spaceplace.nasa.gov/starfinder3/en/). Throughout this book when referring to the timing of a constellation as a backdrop or background to the sun, the actual dates from this table are used.

Table 1.1. The Dates for Constellations on the Ecliptic			
Constellation	Actual Dates for the Zodiac	Days as Background	Traditional Dates Used in Astrology
Pisces	Mar. 12–Apr. 18	38	Feb. 19–Mar. 20
Aries	Apr. 19–May 13	25	Mar. 21–Apr. 19
Taurus	May 14–June 19	37	Apr. 20–May 20
Gemini	June 20–July 20	31	May 21–June 21
Cancer	July 21–Aug. 9	20	June 22–July 22
Leo	Aug. 10–Sept. 15	37	July 23–Aug. 22
Virgo	Sept. 16–Oct. 30	45	Aug. 23–Sept. 22
Libra	Oct. 31–Nov. 22	23	Sept. 23–Oct. 23
Scorpius	Nov. 23–Nov. 29	7	Oct. 24–Nov. 21
Ophiuchus	Nov. 30–Dec. 17	18	
Sagittarius	Dec. 18–Jan. 18	32	Nov. 22–Dec. 21
Capricornus	Jan. 19–Feb. 15	28	Dec. 22–Jan. 19
Aquarius	Feb. 16–Mar. 11	24	Jan. 20–Feb. 18

Despite these differences, stargazing and astrology are not incompatible. While astronomy explores the stars in their current positions, astrology uses a historical system that holds a great deal of meaning. Although I am not an astrologer, I believe that practitioners can draw energy from the constellations that appear in the night sky to enhance their work. After all, observing the stars is a magical experience not to be missed.

While stargazing, if you are not sure whether you are looking at a star or a planet, remember the words to the song "Twinkle, Twinkle, Little Star." Stars twinkle; planets do not. Traveling from a greater distance, the light waves of a star are bent back and forth many more times before our eyes perceive them than light coming from a planet. This bending of light waves in and out causes the twinkling effect. And speaking of light, anyone who has developed the ability to see auras will be familiar with this technique for observing the less-bright stars. The trick is to avert your eyes slightly and not look directly at them; this is called averted vision. This works because we have two types of light-detecting cells in our eyes called cones and rods. Cones help us detect color and are highly concentrated at the center of the eyes. Rods are off to the side and are more sensitive to light. With a little practice, you can become comfortable averting your eyes to see the fainter constellations.

As Earth rotates west to east, stars, planets, and the moon appear to move east to west —as does the sun—giving them the appearance of rising and setting. Of course, observing stars is easier when it is fully dark, but there may be times when you want to witness a star rise. As we know, darkness doesn't occur all at once like an electric light being switched off. In fact, twilight has three defined stages. The first is called "civil twilight," and during this stage only the brightest stars can be seen. You can even read a book outside without the aid of artificial light. During the second stage, called "nautical twilight," the horizon is still visible and the bright stars used for navigation can be seen. The third stage is "astronomical twilight," and it occurs when the horizon can no longer be seen. Finally, when full darkness occurs even faint stars become visible. The three stages of twilight occur in reverse order at sunrise.

While we know that twinkling is a way to tell the difference between a star and a planet, people in ancient times and up through the Middle Ages did not. They did, however, notice a difference in behavior and made intelligent distinctions. In ancient Egypt, stars were called "imperishable stars" and the planets were called "the stars that never rested."[6] In medieval Europe, the stars and planets were called "fixed stars" and "wandering stars," respectively. Fixed stars rose and set as did the sun, but they seemed to stay in the same pattern in relation to other stars. The planets were called wandering stars because their positions changed within a shorter period of time; months or even weeks. Fifteen stars noted by Heinrich Cornelius Agrippa (1486–1535), author of *Three Books of Occult Philosophy,* were considered particularly powerful for magic by medieval astrologers in Europe and the Middle East. More information on these stars is provided in appendix C.

Long before Agrippa pointed out the significance of the fifteen stars, the Persians designated four stars as being particularly powerful and important. Dividing the heavens into quarters, Persian astronomers designated four of the brightest stars as "the royal stars." With one in each quarter, these stars were used as calendar markers for the solstices and equinoxes and were also considered guardians of the cardinal directions.

In addition to constellations and asterisms, we will take a look at a couple of Messier objects. Catalogued by French astronomer Charles Messier (1730–1817), these objects include nebulae, clusters, and galaxies. Tired of the confusion caused by other space objects while comet hunting, Messier cataloged them so he would know what to expect in certain areas of the sky. Messier objects are noted with his name or initial and a number, and sometimes include a constellation name. For example, there is Messier 44, M44, or 44M Cancri. The last designation indicates that it is located within the area of the sky of the Cancer constellation. Others have traditional names, such as the two included in this book: Messier 45 is the Pleiades and Messier 44 is the Beehive.

Shooting stars were a source of awe and mystery in the ancient world. Of course, these are not stars but meteors, which come from the particles that break away from comets and burn up when they enter Earth's atmosphere. In the ancient world, comets and meteors were believed to be omens or responses from deities to earthly events. As signals, they could foretell of approaching doom or they could predict victory in an impending battle. Writings from 1200 BCE Greece indicated that shooting stars were used as oracles to interpret events or a king's actions. The Chinese, Japanese, and Koreans also studied meteor showers for divination purposes. Meteor showers are still a source of awe. Speaking about the Lyrids in a radio interview, Kelly Beatty, senior contributing editor for *Sky and Telescope* magazine, commented, "Meteor showers are truly magical. It's like the universe communicating with us on some sort of basic, primal level. Meteors are the cosmos in action."[7]

Even though meteor showers are named after constellations, they actually have nothing to do with the stars. Most meteor showers occur around the same time each year as the orbits of Earth and a comet intersect. When this occurs, Earth passes through a stream of particles from the comet that burn up when they enter Earth's atmosphere. Unusually large meteors are called fireballs. Because the comet particles are traveling in the same direction, they appear to radiate from the same point and are usually named for the constellation in that part of the sky. Nevertheless, even though meteors are not stars, meteor showers carry a great deal of energy that we can tap into for ritual and magic. Appendix D provides a list of major meteor showers and the dates they can be seen.

Chapter Two

GETTING STARTED STARGAZING

Whether you want to simply read a star map, step outside and find a particular constellation, or spend time outdoors becoming familiar with the layout of the night sky, this chapter provides tips on getting started. But first, let's learn about celestial coordinates, which aid in understanding and reading star maps.

The celestial sphere is a concept that helps us understand the information on star maps. We can visualize it like Earth's atmosphere, a sphere that surrounds our planet, or as two great domed ceilings covering the Northern and Southern Hemispheres. Like the great ceiling in New York City's Grand Central Station, these domes are decorated with the constellations. As Earth rotates, the stars that we see on the celestial sphere appear to move, rising and setting like the sun. Where the northern and southern domes meet is called the celestial equator, which is directly above the physical equator on Earth. As mentioned in chapter 1, the celestial north and south poles are directly over their corresponding points on Earth. Our view of the celestial sphere is defined by the horizon. If we are in flat, open country, the horizon is larger than what is visible in a city. Directly overhead is an imaginary point called the zenith. An imaginary line drawn from north to south through the zenith divides the sky into east and west.

Coordinate points on Earth are called latitude and longitude. On a map of Earth these are indicated by lines that run east and west, and north and south. Latitude lines run east and west parallel to the equator. Marking the distance from the equator, latitude is measured in degrees. You will find more information on latitude in appendix A. Knowing your latitude will help you determine which constellations will be visible to you. The equator is zero latitude. Latitude lines north of the equator are indicated with a plus sign, and those south of the equator with a minus sign. On the celestial sphere, latitude is called "declination" and it is indicated in the same way using plus or minus signs to indicate positions north or south of the celestial equator, which is zero declination.

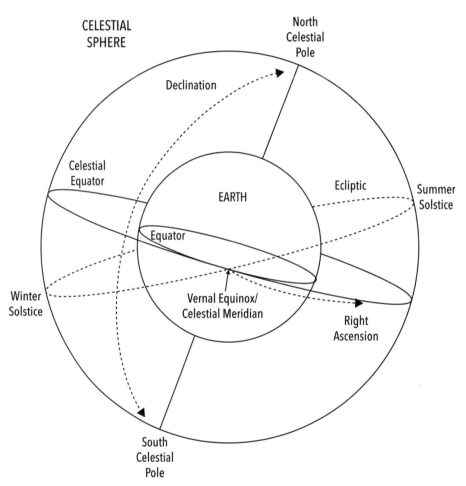

Figure 2.1. The concept of a celestial sphere and celestial coordinates helps us understand the information on star maps and how it relates to Earth.

Longitude on Earth is also measured in degrees from an imaginary line called the prime meridian, which runs from the North Pole to the South Pole. Like the equator, the prime meridian is zero degrees. It is no accident that this line passes through Greenwich, England, the home of the Royal Observatory. Founded in 1675 by King Charles II, the Royal Observatory was instrumental in establishing a common meridian for east/west coordinates and for calculating time worldwide. Time on this meridian is called Greenwich Mean Time.

The equivalent to longitude on the celestial sphere is called "right ascension," and it is measured in hours. Just as our day contains twenty-four hours, right ascension runs eastward from zero to twenty-four. Zero right ascension is the point where the celestial equator and ecliptic cross. Sometimes called the celestial meridian, this point marks the spring equinox.

It is also important to understand star names. Although the traditional names of stars are frequently used, they also have a standardized scientific designation. As noted in chapter 1, astronomer Johann Bayer established a naming convention. This naming system uses the Greek alphabet starting with the brightest star in a constellation. For example, Alpha Centauri is the designation for the brightest star in the constellation Centaurus, and Beta Centauri is the second brightest. These designations use the Latin genitive or possessive form of the constellation's name. For example, Aldebaran is the alpha star in the constellation Taurus, so its Bayer designation is Alpha Tauri, which means "the alpha of Taurus." However, there are exceptions when it comes to the alpha designation, as is the case in the constellation Orion where Beta Orionis is actually brighter than Alpha Orionis. These two stars are better known by their traditional names of Betelgeuse (alpha) and Rigel (beta).

Because of the Bayer naming convention, it is helpful to become familiar with the Greek alphabet as star maps generally use only lowercase Greek letters to distinguish stars rather than spell out their names. Table 2.1 contains the Greek alphabet. The order of letters in this table runs vertically down the columns, i.e., alpha, beta, gamma, and so forth.

Table 2.1. Lowercase Letters of the Greek Alphabet Used in Star Names			
α Alpha	η Eta	ν Nu	τ Tau
β Beta	θ Theta	ξ Xi	υ Upsilon
γ Gamma	ι Iota	ο Omicron	φ Phi
δ Delta	κ Kappa	π Pi	χ Chi
ε Epsilon	λ Lambda	ρ Rho	ψ Psi
ζ Zeta	μ Mu	σ Sigma	ω Omega

To accommodate the limited number of letters in the alphabet for constellations that have more than twenty-four stars, Bayer used lowercase and then uppercase Latin letters. For example, the twenty-sixth brightest star in Centaurus would be called "b Centauri." The forty-ninth star would be "AA Centauri." This cumbersome method of naming stars gave way to Flamsteed numbers.

Developed by English Astronomer John Flamsteed (1646–1719), stars are numbered in each constellation without regard to their brightness starting with the lower, rightmost star on a star map. Actually, stars with a Bayer designation (Greek letter) also have a Flamsteed number. Like the Bayer designations, the Flamsteed system uses the Latin genitive form of the constellation's name. For example, the famous star Betelgeuse (traditional name) in the constellation Orion can be called 19 Orionis (Flamsteed designation) or Alpha Orionis (Bayer designation). You may even find it noted as 19 Alp Ori on star charts. Its Flamsteed number is 19, its Bayer designation is alpha, and Ori is the abbreviation for Orion. Most maps use the Bayer designation for stars that have them and the Flamsteed numbers for those that do not.

In some cases, a star is not just "a" star. It can be a binary, double, or multiple star system, and this is where naming conventions get complicated. You may find designations using numbers, Latin letters, and sometimes superscript numbers and letters. Most often, you will see multiples represented as Alpha-1 and Alpha-2, and so forth, or sometimes Alpha-A and Alpha-B. The difference in using a number or letter is to distinguish multiple stars that are far apart or closer together (relatively speaking), respectively. For our purposes, and simplicity's sake, I have used the naming convention where a number is appended to the star's name. For example, Algorab, the delta star in the constellation Corvus, is a double star with its components designated as Delta Corvi-1 and Delta Corvi-2.

Of course, there are exceptions and the alpha star in Centaurus is one of them. It has three stars, two of which are identified as Alpha Centauri-1 and Alpha Centauri-2. However, instead of being called Alpha Centauri-3, the third star only has its traditional name, Proxima. Another exception is that spectroscopic binary stars do not have completely individual designations. For example, the delta star in Lyra has three components, but two of them share the Delta Lyrae-1 designation as Delta-1a and Delta-1b. The other star in this threesome is called Delta Lyrae-2. I have included this information for those who want to incorporate a full range of star color into their magic. There is more about working with star color in chapter 3 and appendix B.

Before heading outdoors to find your way among the stars, it is helpful to become familiar with a star map. This can be a simple, conventional piece of paper, a more versatile planisphere, or a high-tech smartphone app. Knowing what you are going to be looking at and having something to refer to when you get outside will make star finding easier.

There are two conventions to get used to with a star map. Unlike the terrestrial maps that we are used to, many star maps have east and west reversed; east is on the left and west on the right. These are designed for the user to face south, which puts east and west in their correct orientation. The other convention for star maps is to have east on the right and west on the left where we are used to seeing them, but north and south reversed. These are designed for the user to face north. In both conventions, the direction that a user should face is the one noted at the bottom of the map.

For consistency, all of the maps in this book have been drawn using the first convention of facing south. Likewise, the directions given throughout the book for locating stars and constellations assume that the reader is facing south. It is important to note that the maps in this book are approximations to show the position of constellations and how they relate to each other. Also, they are intended to suggest the constellations rather than reproduce entire star patterns, which makes it easier to use them in ritual and magic work. The star maps that appear with the individual entries in this book show where the suggested star patterns fall within the overall figure of their constellations. Keeping the maps consistent means that some of the graphical depictions of the star figure characters may appear upside down or sideways. Hercules and Pegasus are two examples. While this may seem odd from a pictorial aspect, it is actually helpful to have these views in our minds when we are outside at night looking at the constellations.

A number of star maps in books and on the Internet are drawn in the shape of a circle to simulate the celestial sphere. This type of map is used by holding it over your head and aligning it with the cardinal directions. The center of the star map represents the zenith. Because the stars we can see change with the months and seasons, it is important to have the appropriate star map, as well as one that accommodates your latitude. The Internet is an excellent source for obtaining current star maps. A good source is the Evening Sky Map, which is produced monthly and available at SkyMaps.com. These maps also note the position of the moon throughout the month as well as star magnitudes.

The magnitude of a star indicates its brightness. While the Bayer naming convention also does this, it is somewhat arbitrary and only indicates the order of star brightness within a constellation. Stellar magnitude is applied to all stars regardless of constellation. It began with Greek astronomer Hipparchus around 129 BCE when he designated the brightest stars as being of the first magnitude. He called less bright stars second magnitude, and so on. The faintest stars were sixth magnitude. Ptolemy adopted this system, and it remained unchanged for about fourteen centuries. With the aid of his telescope, Galileo (1564–1642) found many more stars than Hipparchus and Ptolemy had been able to see. Over the centuries, as bigger and better telescopes were developed and more stars could be seen, the magnitude system was expanded. However, it has remained locked into counting backward with the largest numbers representing fainter stars. As light-measuring equipment advanced, the magnitude system was refined with the brightest stars now having negative numbers.

The simple long and short of the magnitude system for laypeople looking at a star map is the size of the dots that represent stars. The larger the dot, the brighter the magnitude of the star, which for most of us is more straightforward. However, now you will know why a map legend will show bigger numbers for smaller dots. The star maps in this book do not show magnitude, so the dots are all the same size.

Unlike the seasonal maps, another type of map called a planisphere is good for the entire year. It is especially helpful for becoming familiar with the night sky through the months and seasons. This is also a good tool if you are an armchair stargazer and simply want to know which constellations are current in the night sky. The planisphere is flexible and can be set to show the sky for any date or time during the year. Like any star map, it is important to get one that is appropriate for your latitude. Planispheres cover a lot of ground, so to speak, and run in ten-degree increments, for example 30° to 40° north.

A planisphere is made of plastic or laminated cardboard and has a center disk that can be rotated. With the constellations printed on it, this center disk represents the sky and shows the rotation of stars around the north celestial pole. Of course, one designed for the Southern Hemisphere rotates around the south celestial pole. The larger, stationary part of the planisphere shows the horizon. Unlike star maps, east and west are in the positions (right and left) that we are used to; however, "north" is printed at the bottom of the horizon, which means you need to face north. While the planisphere is low-tech, it is versatile. If you turn it over, the back has "south" at the bottom horizon so you can use it facing south, too.

To use a planisphere, simply rotate the center disk until the current date and time are aligned. Add an hour if the current time is on daylight-saving time. Of course, being able to rotate the "sky" also allows you to display the constellations for any date and time during the year. In addition, by rotating the disk you can determine when a constellation will rise and set. To do this, rotate the center until the constellation is on the eastern horizon, and then note the time and date. This will tell you when the constellation will rise. Rotating the center again until the constellation is on the western horizon will give you the setting time and date.

Of course, there's an app for that. Smartphone technology puts so much at our fingertips, including the stars. Two popular apps are StarMap, available at iTunes (http://www.apple.com/itunes/) for the iPhone and iPad, and Google Sky Map for Android systems (google.com/mobile/skymap/). A good source for a range of sky apps is AppAdvice (appadvice.com/appguides/show/astronomy-apps). For computers, an online source for star maps is the website of *Sky and Telescope* magazine (www.skyandtelescope.com), which is full of great information, interactive sky charts, and information on space-related apps. For anyone who prefers armchair stargazing or is unable to go out to a good location, apps on smartphones and tablets and interactive websites are a good way to learn about the stars.

Getting Outside

The best place to begin is in your own backyard or neighborhood. Get to know what you can see when you step outside your door. If you live in a city, you may only see the brightest stars, but this is a good place to start. Light pollution is the biggest problem when stargazing in an urban area because the less-bright stars and the delicate band of the Milky Way are obscured. Luckily this is changing as cities and towns are switching to energy-saving downward-directed lighting.

Once you get to know what you can see from your doorstep or backyard, you may want to plan an excursion out to the countryside for a good dark sky. Check the moon phase for the night you plan to go out as the brightness of a full or near full moon can make it difficult to view faint stars. Follow the example of the experts who locate observatories on mountains to get above as much air turbulence and pollution as possible. Even if you cannot go to the mountains, hilltops are better than valleys.

However, if you stay in an urban or suburban area, station yourself as far away as possible from streetlights or other bright lighting. While your eyes will need about ten minutes or more to adjust to the dark, you don't need to wait that long for stargazing. As your eyes adjust, you'll notice more and more stars. Stars that can be seen with the naked eye are called lucid stars.

Whether you are in your backyard or out in the countryside, there are a few things that are useful to have on hand. A blanket is useful because it is often more comfortable (and less strenuous on the neck) to lie on the ground rather than sit in a chair. Along with a star map, take a flashlight but cover the lens with red cellophane so your night vision will not be disrupted. If you go stargazing often, you might consider purchasing a red light flashlight. A compass is a good tool to take along, too, so you can orient your star map correctly.

As we have learned, most stars and constellations appear to rise and set like the sun, but a few do not. These are called circumpolar constellations because they appear to circle the pole. In the Northern Hemisphere, the circumpolar constellations covered in this book include Ursa Major, Ursa Minor, Draco, Cassiopeia, and Cepheus. In the Southern Hemisphere, the circumpolar constellations covered in this book are Crux and Hydrus.

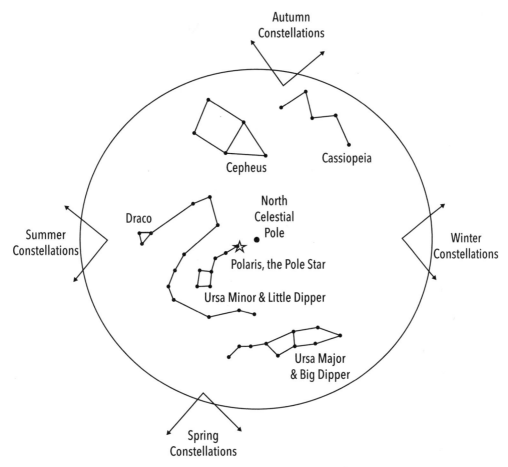

Figure 2.2. The northern circumpolar constellations

Now that you have your flashlight and star map and are comfortable on your blanket or chair, where do you start? First, check your compass to find which way is north, and then orient your star map to it. If you are using a map in this book, north should be to your back so you are facing south. When you look at the sky, the easiest "landmark" (so to speak) in the Northern Hemisphere is the Big Dipper. Since it is part of a circumpolar constellation, it will be visible all year. If you are in the Southern Hemisphere, look for Hydrus.

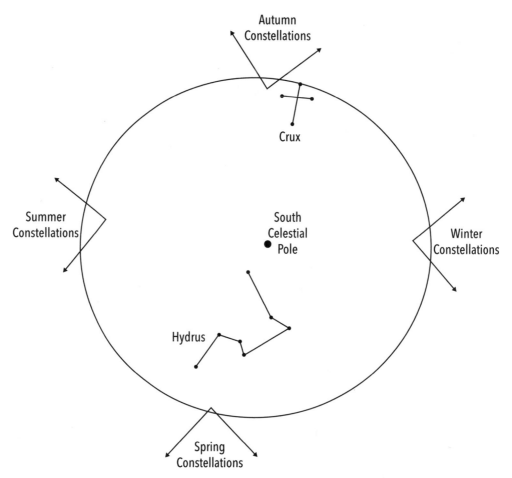

Figure 2.3. The southern circumpolar region

Once you find your circumpolar landmark, locate the brightest stars, which will be your next landmarks. Check your star map for these, choose one, and then find it in the sky. Now that you have found a primary star, check your star map again for others in the same constellation, and then find them in the sky. Take your time and don't rush. It may take a while to get used to this different perspective. When you are ready, work your way around the sky to locate the constellations and stars that interest you.

If you see a star that seems out of place, it is likely to be a planet. Remember, stars twinkle, but planets do not. One last word about stargazing: be sure to dress appropriately for the weather and go where it is safe to be outdoors at night.

Once you find the stars and constellations with which you want to work, set aside the maps and apps and simply be with the cosmos.

In the next chapter, we will learn how to draw on the stars for magic.

Chapter Three

ENGAGING STAR ENERGY
FOR MAGIC AND RITUAL

WORKING WITH ENERGY

Working with star energy is no different from drawing down the moon or other magical energy. The first step is knowing what energy actually feels like. Here's a simple exercise using the palms of the hands that most energy workers start with. Secondary chakras/energy centers are located in the center of the palms. They can be used on their own or in conjunction with the major/primary seven chakras located along the spine and head.

Begin this exercise by sitting comfortably in a place that is quiet and where you will not be disturbed. Close your eyes and spend a minute or two focusing your attention on your breath. This will help quiet your mind and bring your body rhythm into a steady ebb and flow. When you feel relaxed, bring your hands together and rub the palms back and forth over each other until you feel them getting warm. Continue for another moment or two and then separate your hands to about shoulder width apart. Keeping your eyes closed, slowly bring your hands closer together until you can feel a little bit of resistance. When you open your eyes you may be surprised at how far apart your hands are from each other. However, don't be discouraged if your hands are close together or you end up touching palms. It takes time to be able to sense this energy.

Close your eyes and rub your hands together again. This time, rub them a bit longer to activate the palm chakras a little more. When you separate your hands, quickly move them toward each other and then away several times. As you do this, it may feel as though there is a ball between your palms that keeps them from touching. Rub your hands together again and repeat the exercise. Afterward, sit with your hands in your lap with your palms up. Keep your eyes closed and focus your attention on the sensations in your hands. This is what energy feels like. Take a break for a few minutes; get up and stretch or walk around the room.

When you feel ready to continue, sit down again. Place your hands in your lap with your palms up. Close your eyes and focus your attention on your palms. Recall the sensation of energy without rubbing your hands together. When the feeling becomes strong, bring your palms near each other and try to sense the ball of energy again. This part of the exercise may take several attempts and it may not happen in one session. Spend a few minutes each day on this exercise until you can feel the ball of energy without rubbing your hands together. This is how we learn to activate our energy. Once you are able to raise energy at will and feel the ball of energy between your hands, spend a few minutes each day working with it. See how large (hands far apart) you can make it and still hold a definite sensation of energy. Also experiment with how small you can make the ball. This will create an intense little field of energy between your hands.

CHAKRAS

In addition to the seven chakras within the physical body, there are several others that have come to be known as gateway or celestial chakras. These can aid us when working with the energy of the cosmos. Three of these are located above the head and the fourth is within the ground beneath our feet. This "earth star chakra" resides approximately three feet below the ground and connects us with the energy of the earth. As Pagans, we may not have put a name to this connection, but we are very much aware of it when we work with the energy of the natural world as well as when ending rituals.

As for the three chakras above our heads, there is disagreement as to how many, where they are located, and their names. For our purposes, we will use the most common names and details applied to them.

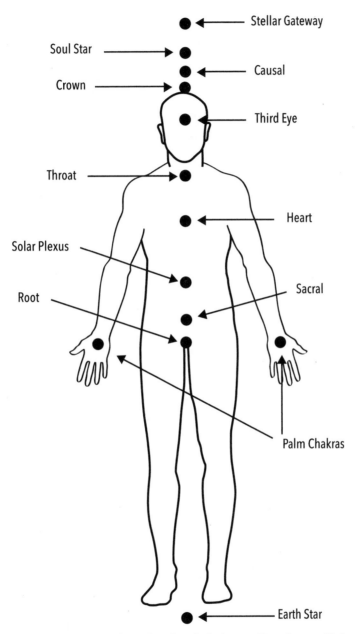

Figure 3.1. The chakras are located within the body as well as above and below it.

The "causal chakra" aligns with the spine about three to four inches above the head. It functions as a portal and regulates the flow of energy from the universal/cosmic level to the personal level, helping to prevent overload. The "soul star chakra" is located about six inches above the head. It is our connection beyond the self to the stars. According to some beliefs, this is the entry/exit point of the soul in the physical body. The third one is called the "stellar gateway chakra." It is approximately twelve inches above the head and provides a multidimensional connection to the cosmos and timelessness as well as to divine energy. These three chakras are instrumental in drawing down and working with energy from the stars.

Table 3.1. Overview of the Chakras		
Chakra	*Location*	*Purpose/Focus*
Stellar Gateway	12 inches above head	Multidimensional connection to the universe
Soul Star	6 inches above head	Connection beyond self to the stars
Causal	3 to 4 inches above head	Regulates flow of energy
Crown	Top of head	Spirituality, awareness
Third Eye	Above and between eyebrows	Intuition, imagination
Throat	Throat	Communication, expression
Heart	Chest	Love, compassion, relationships
Solar Plexus	Stomach	Strength, courage, willpower
Sacral	1 inch below navel	Creativity, sexuality, emotions, passions
Root	Base of spine	Foundation, survival
Earth Star	3 feet below ground	Connection with earth energy
Palm	Center of hands	Energy sensors that support major chakras

Before moving on and working with energy outside of your own, it is important to prepare yourself by being grounded and balanced. As a Reiki practitioner and yoga instructor, I have learned ways to do this simply and effectively. You may hear the terms centered, balanced, and grounded, but what do they mean? Basically they describe a feeling of stability that comes from the core of the body. One of the best ways to engender this experience is through the yoga posture Tadasana, or mountain pose. It is a simple posture, but when it is entered into slowly and mindfully it is very powerful. For this reason, this basic pose never ceases to amaze me. When you become familiar with the energy and sensations of mountain pose, you will be able to evoke the grounding effect without doing the posture.

Start in a comfortable stance with your arms at your sides and your feet about hip-width apart. Hip-width is not the outer sides of the hips; it is measured from the part of the pelvic bone that juts forward. Position your feet in alignment with the pelvic bone so they point straight forward. This brings your big toes slightly closer together than your heels. Already the posture may seem slightly odd because we are not used to standing this way. However, the subtle alignment of this posture opens the body for optimal energy flow.

Continuing with the posture, bring your attention to your feet. Lift your toes and become aware of the soles of your feet on the floor and the stability they provide. Lower your toes to the floor, and then bring your awareness up the body. Your knees should be soft, not locked or bent. Bring your attention up to the front thigh muscles, the quadriceps, and make a tiny adjustment of moving the muscles toward the center of the body. It is a movement that may seem difficult at first, but make it your intention and eventually the physical will follow. This positioning of muscles creates a little space between the legs. Tuck the tailbone slightly, which will flatten the lower back making it easier to hold the posture. Continue to bring your attention up the spine; feel each vertebrae lift. Adjust the shoulders by bringing them up, back, and down so the shoulder blades are closer together. Keep your shoulders relaxed; you are not a cadet at West Point. Let your arms rest comfortably at your sides with fingers together and fingertips pointing toward the floor. Keep your chin parallel to the floor and the crown of your head lifted.

Close your eyes and bring your awareness back down to the soles of your feet. Sense the power of the earth star chakra keeping you connected with Mother Earth. Imagine that you have roots extending from your feet down through the floor, through the earth star chakra, and deep into the earth. Feel the stability of Mother Earth and your connection with her. You are part of her and she is part of you. Now bring your awareness up through your body to the crown of your head. Feel your energy extending up through the causal chakra, into the soul star chakra, and finally the stellar gateway. Sense the lightness of air above you and the vastness of the heavens. Feel it lift you and know that you are part of the stars. Become aware of the difference between these two exquisite energies: one keeping you rooted and stable, the other lifting you to higher purpose. Picture yourself as the fulcrum between these energies as they merge with yours like the delicate kiss of a summer breeze. Hold the posture as long as you can do so comfortably and then slowly bring your awareness back to the room and back into your body. Let go of the sky energy and give yourself a moment or two as you hold the connection with Mother Earth. Slowly release the posture and then move or stretch. Practice this posture every day until you are able to evoke the energy and physical sensations without doing the pose.

In addition to physically sensing energy, engaging the mind with visualization is another component in magic work. Visualization is the use of mental images that run like a cinema in our minds. Daydreams are an example of this; however, in magic it is important to stay for the whole film. In daydreams we may flit back and forth between the images in our minds and the outside world in front of us, but visualization in magic requires focus. A way to achieve this is through the use of multiple senses as well as engaging our emotions. The following is an example to help strengthen your visualization abilities.

Sit comfortably (on the floor if you can) and close your eyes. Imagine that you are at the beach. It's a bright, sunny day and you have just spread a large, soft towel on the warm sand. You notice a conch shell beside your towel and pick it up. Turn the shell over in your hands and examine it. The pale tan outer surface has a rough texture, but the vivid pink opening to the shell's inner spiral is smooth and cool to the touch. The shell is a little heavier than you expected, but you hold it up to your ear to hear the echo of the ocean. After placing it back on the sand, you close your eyes to bask in the heat of the sun and listen to the breaking waves. They begin with a mild thunder and then dissipate into a hiss of roiling water that pushes toward dry sand. The whisper of a salty breeze caresses your skin. A couple of seagulls squabble sharply, then they carry their disagreement farther down the beach. As the noise of the gulls recedes it is replaced by the sound of two children squealing with delight. You open your eyes and watch as they wade into the chilly surf, then laugh and call as they run to their mother. A smile tugs at the corners of your mouth as you soak in the playful spirit of simply being at the beach. Enjoy this feeling for a minute or two, and then gradually bring your attention back to the room where you are sitting. Leave your physical eyes closed for another moment before making the transition to the outer world.

Also try your own visualizations. Imagine the sights, sounds, and smells of a familiar place. Winter holiday gatherings are always a good place to start because they are usually attached to emotions. These exercises provide practice for focusing the mind, emotions, senses, and energy for magical work.

When we draw on star energy or any magical energy, we become a conduit for it as we modify it. We draw it in, shape it with our intentions through visualization, and then send it out with our own energy and willpower. Of course, in order to send it we need to raise our energy level. Chanting, singing, drumming, and dancing are common methods for raising energy, but we can also do it using a more subtle method. You can use the mountain yoga posture or simply come into a straight, comfortable stance. Rub your palms together as in the first energy exercise and when they feel warm, place your left palm on your stomach and your right palm over your heart.

The left hand is over the solar plexus chakra, which is the seat of courage and power. This will activate the energy of this chakra and move it up to the heart chakra, which is being activated with the right hand. The heart chakra is the seat of love and compassion. It serves to moderate the energy of the solar plexus, which our egos can sometimes throw out of whack. With your hands in these positions, feel the energy of both these chakras expanding and merging. Now bring your attention down to your feet and draw earth energy up through your body to the expanding chakra energy. Let it continue to build until you feel that you cannot hold it any longer and then release it. The beauty of this method is that by drawing on earth energy we are tapping into a continuous flow that will enhance but not deplete our own energy. This, of course, also keeps us grounded.

The previous exercise is a particularly good method for working with star energy with only one adjustment. Once the chakra energy is activated and you have pulled earth energy through the earth star chakra into your body, release your hands and turn both palms up toward the sky. The hands can be held in front of the two chakras or at shoulder height. Experiment to find a hand position that is comfortable for you. Just as you drew earth energy up through your feet, draw down the energy of the star or constellation with which you are working. Feel the energy move through the stellar gateway chakra, into the soul star chakra, then through the causal, and into your crown chakra. Pull the energy down through your body. While you are doing this, use visualization to focus on your purpose.

Your hands may begin to tingle as a result of the energy movement. When you feel that you cannot hold the energy any longer, release it as you imagine your intention and energy going with it. This technique can be used to charge a talisman, crystal, or something fashioned for a spell with the star energy you have drawn. Have that object close by so you can pick it up or touch it when you release the energy. Visualize the star energy and your energy going into that object.

ACCESSING STAR ENERGY WITH A SPIRAL

The most common type of galaxy is the spiral, which has long arms that wind toward the center. Our own galaxy, the Milky Way, is a spiral. The spiral is one of the most common motifs in a wide range of cultures from around the world and throughout time. The earliest example consists of a series of seven spirals carved into a mammoth's tusk that was found in Siberia dating to approximately 24,000 BCE. Spirals dating from between 13,000 and 10,000 BCE found in the LaPileta Cave near Gibraltar, Spain, have been interpreted as symbolizing energy and cyclic time.

The spiral motif has been found on Neolithic (New Stone Age) pottery dating to 6300 BCE in southeast Europe as well as in Australia. To the Hopi people of the southwestern United States, it is a symbol of the powerful Earth Mother. A famous spiral carved into a rock in Chaco Canyon, New Mexico, was created by the Anasazi (900–1250 CE). It catches a thin ray of light at sunrise on the winter solstice. Likewise, on the same day of the year at the Newgrange passage tomb in Ireland, a ray of light travels down the sixty-two-foot corridor to illuminate a set of triple spirals on the back wall of the inner chamber.

The spiral is a dynamic symbol of primary life-force energy, which is most evident when it is depicted morphing into plants. We find this from Minoan Crete to high Celtic design, symbolizing growth and transformation of both plants and animals. The spiral held a symbolic role in the myths of people who were in tune with their environments and lived in respectful relation with the natural world. To many ancient people, this moving cosmic energy was evident in the natural world in the form of meandering streams, the curling ocean wave, and the eddy of a whirlpool. Representing the birth of a star in the swirling nebula of space, the spiral is "symbolic of the mystic center and the unfolding of creation." [8]

This leads us to our application of a spiral that can be used like a labyrinth, the design of which is believed to have evolved from the spiral. According to Dr. Lauren Artress, founder of Veriditas, a worldwide labyrinth project, a labyrinth functions like a spiral, creating a vortex at its center. [9] This may be due to the movement of energy because one half of the walk moves clockwise and the other half moves counterclockwise. I have found that walking a spiral can provide just as profound an experience as walking a labyrinth.

When we walk a spiral, going toward the center is symbolic of moving inward toward the self as well as moving toward primal source—star stuff. Walking a spiral is both a descent and ascent that provides an opportunity to find balance and wisdom. Each journey on a spiral provides a unique opportunity for us to gain personal insight and connect with star energy. A spiral walk is an excellent method to prepare for ritual, astral travel, divination, magic, or any type of psychic work.

To make a spiral, clear as much floor space as you can, or if you have privacy you can create one in your backyard. The easiest thing to use to make a spiral is a couple of strands of Christmas garland. Occasionally this type of garland may include star shapes, which is perfect but not necessary. Rope also works well. Begin laying out the garland or rope at the center of the spiral and leave enough space to sit. Make the space within each wind of the spiral wide enough to form a path for walking. Use this setup time to focus on your purpose and to quiet your mind.

When the spiral is finished and you are ready to walk it, stand at the entrance for a moment or two with your eyes closed to further focus the mind before stepping onto the path. When we walk a spiral, we move slowly and with intention. It is important to move at your own pace, and if you feel the need to pause during the walk, do so. The walk itself consists of three parts: preparation, intention, and integration.

"Preparation" is the walk into the center. This provides time to release attitudes or emotions that may inhibit us as well as time to fully focus on our purpose. We are spiraling inward and releasing the ego's control as well as anything that may hold us back. The walk inward toward inner self also prepares us to reach outward to connect with the cosmos.

The "intention" part of the walk occurs when we come to stillness at the center of the spiral. After sitting down, it is helpful to activate the palm chakras and place them on the solar plexus and heart center as you draw down star energy. Take time to sit with this energy to experience it and know it with your body. Let your intuition guide you for the appropriate time to bring this part of your walk to a close. Release the star energy, and then pause for a moment before initiating your walk out.

"Integration" takes place as we wind our way out of the spiral and begin to assimilate any insights and experiences that may have occurred while sitting in the center. The power of the inward and outward movement may make us feel extremely grounded or even a little wobbly from the energy that has been generated. Upon emerging from the spiral, it is helpful to take a moment or two to stand or sit in silence.

If you want to charge an object with star energy, carry it with you or place it at the center of the spiral when you set it up. After you walk to the center and begin to draw down star energy, hold or place your hands on the object and envision the energy passing through you into it. Take the object with you when you walk out or leave it in place for a while in its own swirling spiral galaxy of energy.

How to Use Star Magic

As mentioned in the above exercises, we can draw star energy into a talisman, crystal, or something fashioned for a spell to boost the energy of our intentions. Additionally, a cord that you want to use as a special belt for ritual, divination, or other practices can be prepared by drawing star energy into it and making a knot for each of the stars in a particular constellation. To give the belt a little more punch, the color of the strands within the cord can be coordinated with the colors of the stars in a constellation. We can also draw down star energy for use in ritual and, quite appropriately, before astrology work. While specific suggestions are provided for each constellation, these are by no means the only ways that star energy can be used. Let your creativity and intuition guide you to find uses that you can tailor to your personal situation. In some of my suggestions I have provided specific words, but don't hesitate to use your own.

One of the items I frequently mention for working with star energy is the star-shaped glitter/confetti that is sold at many arts and crafts stores. While this is available in various sizes, the larger ones are easier to work with when laying out pieces to create a star pattern. Laying out the pattern of a constellation on your altar is one way to incorporate star energy into your rituals or other activity. Each constellation entry in the following chapters includes a picture of its star pattern. For example, to draw down the energy of Corvus the Crow lay out six pieces of glitter/confetti, one for each star, to match the pattern in the picture. As an alternative to the glitter/confetti use gemstones, flowers, or any other object that seems appropriate to mark the position of each star. Also, if you are drawing on a constellation's energy to aid your garden for growth in the spring or to protect your property, natural objects can remain in place to hold the energy and strength of your purpose.

Candles are an important component for ritual and spellwork, and they are especially apropos for star magic because we do a great deal of it at night. There are several ways to employ candles, one of which is through color. When it comes to stars, most may appear white or bluish-white when viewed without the aid of binoculars or a telescope, so it may be a surprise that some stars are rather colorful. Actually, the color of some of the brightest stars can be seen with the naked eye. A star's color is related to its surface temperature, and according to scientists, white is hotter than red, and blue is hotter than white. The most common colors found in stellar classification are as follows: Class O stars are blue, B stars are blue-white, A stars are white, F are yellow-white, G are yellow, K stars are orange, and M stars are red. Less common are the purple and violet stars. Green or a greenish tint is considered fairly rare.

A simple way to amplify star energy is to coordinate its color with that of a candle, gemstone, or other object that you may use when laying out the star pattern of a constellation. For example, you may want to use something red when drawing on the energy of the red star La Superba in the Canes Venatici constellation. Red helps to support intentions such as love and passion, quite obviously, but this color also aids in breaking hexes, building defensive magic, finding guidance, and acquiring wisdom. I have included color along with other information about stars within the information for their respective constellations. For convenience, appendix B lists the colors of all the stars mentioned in this book.

As we have already learned, a star is not always just "a" star. This means that without a telescope it may appear as a single star when it is actually more than one. It may be a binary star, two stars that revolve around each other, or groups of stars. Where possible, I have included color information for each star in these multiples, which can provide flexibility for color magic because one "star" can have several colors. For example, Algorab, the delta star in Corvus the Crow, has two components one of which is yellow-white and the other a lilac/purple. On the other hand if you want to keep it simple, use the color of the primary star. For example, Regulus, the alpha star in Leo, has three components and each is a different color. You can use one gemstone (or other object) in the color of Alpha-1, which is blue-white.

Whether or not you decide to coordinate star and candle color, you can still combine candle and star magic by drawing the star pattern of a constellation on your candle. There are two ways to do this. One is to use a fine-point, felt-tip pen to draw on the candle surface. The other is to use a large sewing needle (or other finely pointed object) to scratch the pattern into the candle. Use the graphics provided with each constellation for the basic star pattern. You can simply use dots to indicate each star and its place in a constellation or connect the dots as in the graphics. If you are working with one particular star, you can draw the whole constellation and make that star larger than the others or simply write the name of the star on the candle.

Because our personal energy is vital to working magic or engaging in psychic work, aligning the chakras and balancing the aura can significantly enhance our experiences and aid in developing abilities. To do this, sit comfortably and follow the suggestions to ground and center your energy, or use any method that works best for you. When you feel ready, lift your hands (palms upward) to shoulder height and visualize the energy of the stars entering through your hands. See this as pure white light moving down to your root chakra located at the base of the spine. As the chakra energy is activated, it glows a bright red color and radiates upward. Visualize the chakra energy and the white light of the stars moving up to the sacral chakra located an inch or so below the navel. As it becomes activated, it glows orange and radiates upward. Visualize the chakra energy and the white light of the stars moving up to the solar plexus chakra, which will radiate a yellow light. Continue up through the remaining chakras: The heart chakra radiates a green color, the throat chakra radiates blue, the third eye chakra located between and slightly above the eyebrows radiates indigo, and the crown chakra at the top of the head radiates violet.

From the crown chakra, think of the energy radiating a rainbow of colors out from the top of your head, up through the three celestial chakras, and then down over your body. Visualize these colors being replaced with pure white star light coming down through the stellar gateway chakra, surrounding you and washing your aura, the energy field that extends out from the body. With your hands a few inches from your body, act as though you are raking your fingers through your hair. Start at the crown of your head and work downward. Continue this raking motion down over your entire body. When you get to the area of your feet, use a sweeping motion to whisk any negative energy away from you. Sit for a moment with your hands in your lap, palms up, to allow the energy to settle. You have just completely activated, cleansed, and balanced the energy within and around your body. Magic or psychic work that you engage in after this process will not be hindered by the little negativities that crop up in daily life and sometimes stay with us.

ASTRAL TRAVEL, DREAM WORK, AND RITUAL

What could be more perfect than using the power of the constellations to boost astral travel? According to author and scholar John Michael Greer, the astral plane is important because it is where "most magical energies come into manifestation."[10] In addition, some believe that the astral plane is where deities and spirits reside.

Our bodies consist of layers: the physical, etheric, and astral. The astral body, sometimes called the "star body," is often described as shimmering as though it consists of a multitude of tiny stars. There is disagreement as to whether the aura is part of the astral or etheric bodies or both. At any rate, the astral body is our vehicle of consciousness, and astral travel is generally defined as any state in which the consciousness is immersed somewhere different from the physical body.

Astral travel has also been referred to as a soul flight, dimensional journey, psychic travel, and astral shifts. Some people consider these as very similar and others make distinctions among them. However, the common point is that the consciousness goes somewhere out of the physical body. Whether it goes to the astral plane, the underworld, or the fairy realm, the stars can aid us with energy for travel and protection. When I have journeyed, I usually have something like a wand or crystal on my physical body that the astral body will have, too. Using methods previously described, I draw protective energy into these objects. When I'm ready to travel, I raise energy in the physical body and then visualize it rippling out through the etheric and astral bodies. As I see the stars in my astral field twinkling, I visualize energy coming from the constellation or star I've chosen to work with and transporting me to my destination. Support from the stars also helps to enhance contact with power animals, spirit guides, and astral familiars that we meet during these travels. This method for initiating astral travel can also be used to prepare for divination or any type of psychic work.

Because most of us sleep at night when the constellations are visible above, star magic is an aid to dream work. In the fertility of darkness, stars help us kindle the spark that will guide us on new paths, open our creative channels, and illuminate the areas of our lives that we may not be able to access during waking hours. Unlike other practices for which we raise energy, in dream work we want to avoid too much active energy that will keep us from reaching a deep sleep. This is where the use of an object such as a dream pillow is instrumental. It can be imbued with star energy ahead of time using the technique described for talismans. In place of a dream pillow, draw a constellation star pattern on a piece of paper or cloth that you can slip under your pillow. If you choose to work with an individual star, you can write the name of the star and coordinate the color of the cloth with the color of the star. In addition to writing the name of a star or drawing a constellation on a piece of paper, also write down your intention for the dream work. On the evening after your dream work, create sacred space and burn the paper; ideally outdoors under the stars or wherever it can be done safely.

The fixed stars used in medieval magic were associated with particular plants into which the star's energy is easily drawn. If you are working with one of these fifteen fixed stars and making a small dream pillow, stuff it with leaves or flowers of the dried plant or plants that correspond with the star. For extra measure, also coordinate the color of the pillow fabric with the star color and sew or draw the name of the star on the outside of the pillow.

In ritual, the directions are often called in as guardians and in this respect the ancient Persian royal stars fit well. The star Aldebaran in Taurus is the Guardian of the East, Fomalhaut in Piscis Austrinus is the Guardian of the South, Antares in Scorpius is the Guardian of the West, and Regulus in Leo is the Guardian of the North. All of these are alpha stars in their respective constellations. An example of using a star to call in a direction would be: *"We call to the Guardian of the East, Aldebaran, brightest star of Taurus, bring the power of air as you send forth your light from on high."*

Because a star's color can be an important feature for its use, you may want to incorporate it into whatever you say when drawing down a star's energy. Refer to appendix B for a detailed list of stars and their colors. Using La Superba in the Canes Venatici constellation as an example: *"La Superba, red and bright; Send energy to this circle tonight. Amongst the stars in the dogs of spring; La Superba, the power of fire bring."*

An alternative to using the royal stars or star color for calling in the directions is to select four stars or constellations that will be in the sky during the time of your ritual. Throughout the book you'll find suggestions on how the energy of various constellations can be relevant to the sabbats. However, don't hesitate to use others if you feel that they are more appropriate to your ritual and magic work. Also consider the circumpolar constellations, those that never rise or set: Cassiopeia, Cepheus, Ursa Major and Minor, and Draco. I like to think of Cepheus and Cassiopeia as the Lord and Lady who watch over my rituals throughout the year. More on these later in the book.

And now, let's leave our backyard of the solar system and explore beyond to the stars.

Chapter Four

THE SPRING QUARTER OF MARCH, APRIL, MAY

As the earth awakens and loosens the bonds of winter, Imbolc's promise of renewal is carried forward. Even where I live in northern New England, March ushers in a freshness in the air that lightens the heart and entices us to spend more time outdoors. This windy month blows away the staleness of winter and symbolically helps us cast away the things we want to remove from our lives to make way for growth. In the past, this part of the year was called the Plow Quarter because it signaled the time to ready the fields for sowing and planting.

Spring arrives on Ostara and the natural world seems to vibrate with quickening energy. After each rain shower, life bursts forth with renewed vigor turning the landscape many shades of lush green. The softness of spring is like a shimmering pool of life-force energy that feeds our bodies and souls. Plants, new life in the womb, and the creativity of our spirits blossom and grow, leading us into the magical month of May and Beltane. Warm weather entices us to walk barefoot in the grass and enjoy nature's potpourri of fragrances carried on the gentle kiss of a breeze. Under the stars, it is easy to feel the greatness and magic of the universe and the enchantment of the fairy realm. Open your heart and let your energy reach out to the constellations.

The spring night sky is populated by people, animals, and objects: a woman, a herdsman, a centaur, hunting dogs, bears, a lion, a crow, a water snake, a cup, weighing scales, and a crown. Some of their ancient stories overlap as do their significance for twenty-first-century Pagans and Wiccans. We will take a look at the constellations, their stories, and their stars as we discover new ways to draw on their magical influences.

Some of the bordering constellations noted throughout this chapter may fall within other seasons. The directions given to locate constellations and stars assume that the reader is facing south. Also, have a look at chapter 8 as some of the southern constellations listed only in that chapter may be visible to you.

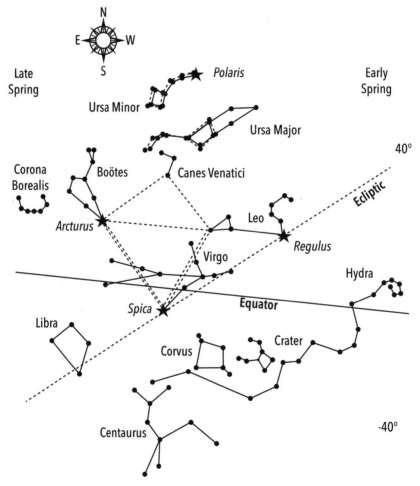

Figure 4.1. The spring sky. The dotted shapes show the four asterisms of the Great Diamond, the Spring Triangle, and the Big and Little Dippers. The four brightest stars are also noted.

THE SPRING CONSTELLATIONS

Boötes: The Plowman/The Herdsman, Green Man

Pronunciations: Boötes (bow-OWE-teez); Boötis (bow-OWE-tiss)

Visible Latitudes: 90° North to 50° South

Constellation Abbreviation: Boo

Bordering Constellations: Canes Venatici, Corona Borealis, Draco, Hercules, Serpens, Ursa Major, Virgo

Description: The central part of this constellation has a kite shape with a triangle of stars above. The triangle represents Boötes's raised left hand holding the leash of his dogs represented by the constellation Canes Venatici.

To Find: Follow the handle of the Big Dipper to the southeast and according to the saying, "arc down to Arcturus," the bright star located at the bottom of the kite shape.

Considered a plowman or herdsman, one theory about the name Boötes is that it comes from a Greek word that means "noisy" or "clamorous," which describes how shepherds sometime use sound to get their animals moving.[11] Boötes has also been considered a herdsman or wagoner with the stars of the Big Dipper representing his oxen and wagon. He later became known as Arctophylax from the Greek meaning "bear keeper" or "bear watcher" as he appears to follow the Great Bear across the sky.

Polish astronomer Johannes Hevelius (1611–1687) put forth the idea that Boötes was a hunter. He depicted some stars as Boötes's hunting dogs pursuing the Great Bear. This is how the constellation is usually regarded today.

According to legend, Boötes was the son of Demeter and a mortal man. When Boötes was getting on in years, rather than watch him die Demeter gave her son the ultimate reward and placed him among the stars. In some stories, Boötes is attributed with the invention of the plow. Other legends equate Boötes with Icarius (not to be confused with Icarus, who flew too close to the sun), whom Dionysus taught to grow grapes and make wine.

Ptolemy equated the energy of Boötes with Mercury and Saturn.

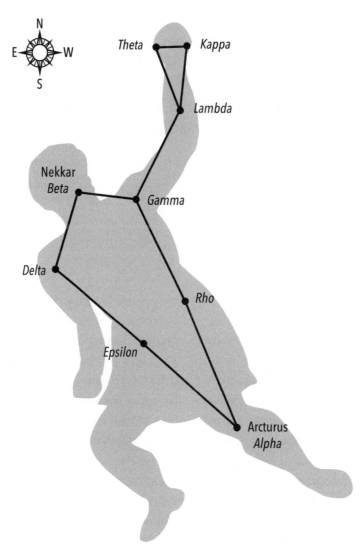

Figure 4.2. Boötes the herdsman, plowman,
Green Man brings us closer to the natural world.

Notable Stars in Boötes

Official Designation: Alpha Boötis

Traditional Name: Arcturus

Pronunciation: ark-TOOR-uhs

This orange star is the fourth brightest in the entire sky. It is located at the bottom of the kite shape and represents the herdsman's left knee. The Arabs called this star the Keeper of Heaven, and the Chaldeans called it the Guardian Messenger. Associated with protection and guidance, Arcturus was one of the fifteen important fixed stars in medieval magic. Arcturus is the eastern anchor point for the Spring Triangle and the Great Diamond asterisms.

Official Designation: Beta Boötis

Traditional Name: Nekkar

Pronunciation: NECK-ahr

Nekkar is a yellow star located at the peak of the kite shape that marks a point near Boötes's left ear. The name of this star comes from Arabic and means "herdsman" or "ox driver."

Magical Interpretations and Uses for Boötes

Although the figure of Boötes does not have horns, he is evocative of the Horned God and is a deity of shepherds and farmers. He represents Pan, Faunus, Cernunnos, Herne the Hunter, and Silvanus. He is lord of the animals and god of forests, fields, and herding. To farmers and herders, the cross-quarter days of the year were more important than the solstices and equinoxes because they marked seasonal changes. Symbolizing fertility, Boötes represents the Lord at Beltane.

Prepare for spring planting and call on Boötes as you take a stick to break up large clods of dirt in your garden. Do this gently with reverence and respect. Take time to pause and smell the richness of the soil. Using your hands, create a level area and draw the simple kitelike shape of the constellation as you say: *"Boötes, Boötes, send forth your light and make ready this ground to receive the blessed seed. So mote it be."*

Whether Boötes is considered as a plowman, shepherd, herdsman, or hunter, he represents someone intimately associated with the land and animals. As such he serves as the

Green Man, the wild and potent spirit of nature, and he can help us connect with the natural world on a deep and meaningful level. Boötes also serves to remind us to observe and know the natural world.

Go outside on a warm spring night and sit quietly. Smell the air, the plants, and the soil. Ask the Green Man/Boötes to reveal something relevant to you. If you think of him as being more of a herdsman, listen for the sounds of animals. Don't look for them or approach them, just close your eyes and listen. Become familiar with night sounds. Also call on the power of Boötes for inspiration in helping the environment or learning about the natural world. Your work or study does not have to take place at night, but seeking his help works best when Boötes is overhead.

Throughout the centuries, this constellation was considered a weather maker. Roman naturalist and author Pliny the Elder (Gaius Plinius Secundus, 23–79 CE) noted that this constellation's rising at dusk "portended great tempests" in the spring rainy season that could bring violet storms.[12] If you engage in weather magic, call on Boötes to aid you. Use an oak or hazel wand to point to each cardinal direction as you say: *"Rain, rain, go away; Come again another day. Bring skies that are fair and blue; Boötes for this I call on you."*

Canes Venatici: The Hunting Dogs/Energy of Renewal
Pronunciations: Canes Venatici (KAN-es veh-NAHT-ih-see);

    Canum Venaticorum (KAY-num veh-NAHT-ih-KOR-um)

Visible Latitudes: 90° North to 40° South

Constellation Abbreviation: CVn

Bordering Constellations: Boötes, Ursa Major

Description: Canes Venatici is a constellation without a memorable pattern, and it is usually noted on star charts by two or three of its brightest stars.

To Find: Canes Venatici is a dim constellation located between Boötes and Ursa Major, slightly closer to and under the bear's tail. Draw an imaginary line from the star at the end of the Great Bear's tail to the star at the end of Leo the Lion's tail. Canes Venatici is about one third of the distance from the Great Bear.

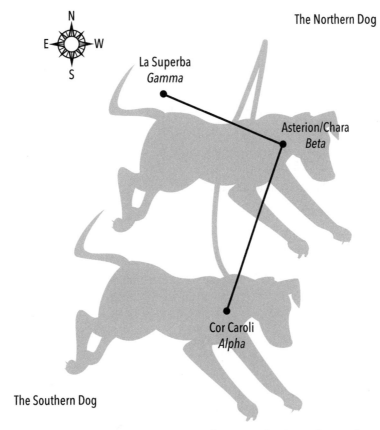

The Northern Dog

La Superba
Gamma

Asterion/Chara
Beta

Cor Caroli
Alpha

The Southern Dog

Figure 4.3. Canes Venatici represents the energy of spring and renewal.

As previously mentioned, this constellation represents the leashed hounds of Boötes the Herdsman. Originally these stars were split between Ursa Major and Boötes with some of them representing Boötes's club. Mistakes in translating Ptolemy's work from Greek to Arabic and then into Medieval Latin resulted in the word "club" ending up as "dog." Boötes was first depicted accompanied by dogs on the 1533 star map of German astronomer Peter Apian (1495–1552). By the seventeenth century, the dogs had acquired names. Asterion, meaning "little star" or "starry" in Greek, marked the northern dog, and Chara, meaning "joy," the southern dog. Over time, the name Chara was also adopted for the name of the northern dog, which tends to cause some confusion.

Previously, these stars were part of a constellation called Cor Caroli, "the Heart of Charles." It was named in honor of King Charles I, who was executed in 1649 during England's Civil War. The constellation was also called Cor Caroli Regis Martyris, "the Heart of Charles the Martyred King," and first appeared on star maps in 1673. Cor Caroli is now used as the name of the southern dog.

In medieval times, the influence of this constellation was believed to bestow a keen mind as well as faithfulness.

Notable Stars in Canes Venatici

Official Designation: Alpha Canum Venaticorum

Traditional Name: Cor Caroli

Pronunciation: core ka-ROLE-ee

Cor Caroli is a double star located on the neck of the southern dog. Alpha-1 is a yellow star and Alpha-2 is white. Cor Caroli is the northern anchor point for the Great Diamond asterism.

Official Designation: Beta Canum Venaticorum

Traditional Names: Asterion; Chara

Pronunciations: as-TEE-ree-on; CHAH-ruh

Asterion/Chara is a yellow star that marks the northern dog. Although it causes confusion, this star of the northern dog is known by two names, one of which was the former name of the southern dog.

Official Designation: Gamma Canum Venaticorum

Traditional Name: La Superba

Pronunciation: LAH sue-PER-bah

La Superba is one of the reddest stars in the sky. It was named the Superb One by Italian astronomer Pierto Angelo Secchi (1818–1878) because of its striking color and its change in brightness, which occurs every 160 days. This star is not used in depictions of the dogs.

Magical Interpretations and Uses for Canes Venatici

Like a number of other spring constellations, the dogs have a dual nature. Throughout time, the dog has been portrayed as a dangerous creature of the night and of the otherworld as well as a faithful companion, guardian, and guide. The dog was associated with themes of spring in the ancient cultures of Europe. According to Lithuanian professor and archaeologist Marija Gimbutas (1921–1994), dogs represented the energy of spring and renewal. Associated with life and death, dogs were the guardians of life who oversaw the growth of vegetation and the fallow periods of rest. According to Gimbutas, "they promote the lunar cycle and plant growth" and functioned as a symbol of becoming.[13]

Canes Venatici can be a powerful influence for your garden, especially a vegetable garden, where the dog's association with fertility and abundance would be apropos. An outdoor statue of a dog would work well to draw on the energy of this constellation. We can take a cue from the ancients who portrayed processions of dogs on their pottery to encourage the growth of plants. Don't worry about being an artist as you can draw or paint the most simple side view to represent dogs end to end (snout to tail) to depict a procession. As an alternative, you could draw two dogs to represent Asterion and Cor Caroli. The image does not have to be large; in fact, I like to draw them on a few bricks that I place around the garden in early spring.

At Beltane, show the fairies that you honor their legendary Cu Sith, the green fairy dog, by placing a green toy dog near flowers in your garden or inside your home. Because dogs were believed to possess supernatural powers including shape-shifting, call on Canes Venatici for aid in your divination practices or other psychic endeavors. In addition, because dogs are well known for their guidance and protection, Canes Venatici can be called upon for protection while traveling or for guidance when journeying. To foster the energy of this constellation during all of these activities, wear a dog collar around your ankle or wrist.

In classic Mediterranean and Celtic myth, dogs were associated with the power of healing, perhaps because of the effect of their self-healing licking. To enlist their magical support for someone you love, write the name of the person on a piece of paper. Fold it up as small as you can and say three times (filling in the blank on the last line with the name of the person to whom you want to direct healing energy): "*Star dogs, star dogs, in the spring; Shine on high and healing bring. Loyal dogs, Canes Venatici; Help bring health to [name] for me.*"

Do this on a clear, starry night and then place the paper on a windowsill or outside. The next day bury it in the ground.

Centaurus: The Centaur/Chiron the Healer

Pronunciations: Centaurus (sen-TOR-us); Centauri (sen-TOR-ee)

Visible Latitudes: 25° North to 90° South

Constellation Abbreviation: Cen

Bordering Constellations: Crux, Hydra, Lupus

Description: The centaur's front legs are marked by two of the brightest stars in the southern sky, Alpha and Beta Centauri.

To Find: Although this constellation can be seen across the southern United States, it is below the horizon for many of us in the Northern Hemisphere. Find the bright star Spica in Virgo and draw an imaginary line south beyond the tail of Hydra the Water Snake. The next constellation to the south is Centaurus.

This constellation was associated with the mythical half-man, half-horse centaur. Sources differ on which centaur the constellation represents, but it is usually considered to be Chiron, who was a mentor to Achilles, Hercules, and Asclepius. Chiron was said to be wise and gentle, unlike the wild and less-civilized centaur Sagittarius.

According to legend, Chiron was the son of Cronus, the Titan god of time, and the sea nymph Philyra. When Cronus's dalliance with Philyra was almost discovered by his wife Rhea, he turned himself into a horse. The resulting child of his union with Philyra was a hybrid of human and horse. As a teacher, Chiron was well known for his knowledge of medicine, music, and hunting. Ironically, he was hit by one of his former pupil's (Hercules) arrows that had been poisoned with the blood of Hydra the Water Snake. There was no cure and because Chiron was immortal (as the son of time), he suffered rather than died. Zeus came to the rescue, releasing him from his earthly bonds and placing him in the heavens.

In other Greek legends, Centaurus taught the Argonauts how to read the sky and placed an image of himself among the stars to guide them across the sea. In addition, Centaurus was often depicted on star maps sacrificing Lupus the Wolf on Ara, the constellation that represents an altar.

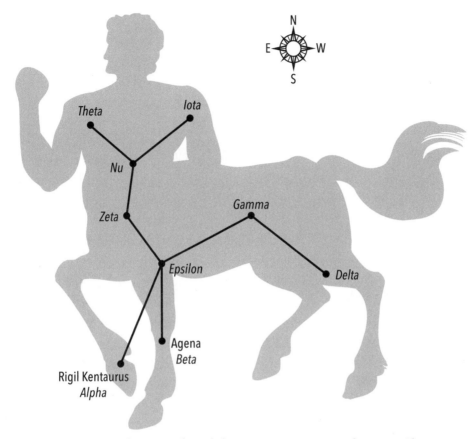

*Figure 4.4. According to Greek Mythology, Centaurus represents the centaur Chiron,
a healer and teacher.*

Located in the Southern Hemisphere, Centaurus was catalogued by Ptolemy in the second century, but mentioned in earlier writings of astronomer Eudoxus. The constellation
was also called Hippocrator, which means "the ruler of horses." The Babylonians equated
this constellation with a Bison-man that was often depicted with four legs like a centaur and
sometimes as a Minotaur with two animal (instead of human) legs.

Ptolemy equated the stars in the human part of this constellation with the energy
of Venus and Mercury, and those in the animal part with Venus and Jupiter. In medieval
medicine, Centaurus was believed to foster good health and long life.

Notable Stars in Centaurus

Official Designation: Alpha Centauri

Traditional Names: Rigil Kentaurus; Toliman

Pronunciations: RYE-jill ken-TAW-rus; toll-ee-MAN

Alpha-3 Traditional Name: Proxima

Pronunciation: PROK-sa-muh

Alpha Centauri is one of the brightest and most famous stars. It is one of the few that is known most widely by its official designation. After the sun in our solar system, Alpha Centauri is the nearest star to Earth. It is actually a multiple star system with a pair of similar stars. Alpha Centauri-1 is a yellow-white star and Alpha-2 is orange. The third star in Alpha Centauri is a red dwarf called Proxima. The name Rigil Kentaurus, which is often shortened to Rigil Kent, comes from Arabic and means "the foot of the centaur." The name Toliman also comes from Arabic and means "hereafter." Alpha Centauri has also been known as Bungula, a name that comes from a Latin word meaning "hoof." This star supports psychic abilities.

Official Designation: Beta Centauri

Traditional Names: Agena; Hadar

Pronunciations: ah-JEEN-ah; hah-DAHR

This blue-white star is located on the leg of the centaur. Its traditional name, Agena, is Latin for "knee." The name *Hadar* comes from Arabic and means "soil" or "earth." This star is associated with health, honor, and friendship.

Magical Interpretations and Uses for Centaurus

Like many of the spring constellations, Centaurus presents us with duality and balance. Here we are dealing with the duality of being half-human and half-animal and the struggle in bringing these two natures together. When we find ourselves struggling with two aspects of self, we can call on Centaurus for guidance or to simply assist us with coming into balance.

Since Centaurus is most often equated with Chiron, we will focus on his qualities, most notably that of a healer. For help in drawing down the energy of Centaurus for healing, lay out the constellation's star pattern using pieces of amethyst, clear quartz, moonstone, hematite, or moss agate. In addition to the crystals and gemstones, have a small bowl of spring water or if you live near the ocean, seawater. Hold the bowl between your hands as you say: *"Centaurus, I call on you to send forth your light to charge this water with healing energy."*

Visualize the energy of Centaurus coming through your celestial chakras, into your body, and into the water. Dip a finger into the water and then draw the Centaurus pattern of stars on the part of your body that needs healing. If this is not easy to do, draw the pattern on the palm of your hand and visualize the healing energy flowing to the area of your body that needs it. When you are finished with your visualization, take the bowl outside and pour the remaining water on the ground as you say: *"Centaurus, may you always be strong, and may your stars shine forever. Blessed be."*

According to some legends, Chiron taught music and other skills to the children of the gods. In addition, he has been associated with psychic abilities and serving community needs. We can call on Centaurus to provide us with wisdom as we embark on these endeavors with this incantation: *"Chiron the centaur, shining above; Send me your wisdom, healing, and love. May this year unfold like music so sweet; I bid honors to you. Merry meet."*

To honor Centaurus or for aid in drawing down his power, plant the flowering herb known as Centaury or Common Centaury (*Centaurium erythraea*) in your garden. In the Middle Ages this plant was used as a cure-all. It was believed to have the power to attract luck and to ward off evil. Its folk names include Chironia and Centaur's Hoof. If you don't have this plant in your garden, you might find it in a meadow or woodland. Wherever you find it, pick three small flowering sprigs for your altar. Spread them fanlike with their stems touching and flowers toward you to represent the energy of Alpha Centauri. Because it is a triple star it carries the symbolism of the sacred number three and can be used to boost magical and spiritual energy.

Corona Borealis: The Northern Crown/Guidance and Power

Pronunciations: Corona Borealis (kuh-ROE-nuh bore-ee-AL-iss);

   Coronae Borealis (kuh-ROE-nye bore-ee-AL-iss)

Visible Latitudes: 90° North to 50° South

Constellation Abbreviation: CrB

Bordering Constellations: Boötes, Hercules, Serpens

Description: Seven of the most visible stars are fairly evenly spaced and form a semicircle.

To Find: Draw an imaginary line east and slightly south from the end of the Big Dipper's handle through the top of the kite shape in Boötes. Look for the distinctive semicircle of Corona Borealis to the east of Boötes.

In Latin, Corona Borealis means "the crown of the north." This name was translated from the Greek *korone*, "crown," and combined with *Boreas*, the name of the north wind. Corona Borealis has also been translated as "the crown of the north wind." This constellation was first catalogued by Ptolemy and known simply as Corona. The ancient Greeks called it the Wreath as well as Ariadnaea Corona, "Ariadne's Tiara." It was also called the Coiled Hair of Ariadne. Most often it represented the crown of Ariadne, the daughter of King Minos of Crete.

According to myth, Ariadne helped the hero Theseus find his way out of the maze after he killed the Minotaur. The Minotaur was a creature with a human body but the head of a bull. Ariadne fell in love with Theseus and gave him a ball of thread to take with him into the maze so he could find his way out again. He promised to take her with him when he left Crete, but instead he abandoned her. She wasn't alone for long, as Dionysus began courting her. This constellation's circlet of stars represents the crown she wore on her wedding day.

The Romans also regarded this constellation as a crown and called it the Crown of Vulcan. They also called it the Crown of Amphitrite, a sea goddess, because of its proximity to Delphinus, the dolphin constellation. At various times, Corona Borealis was considered part of the Boötes constellation, representing a sickle in his hand.

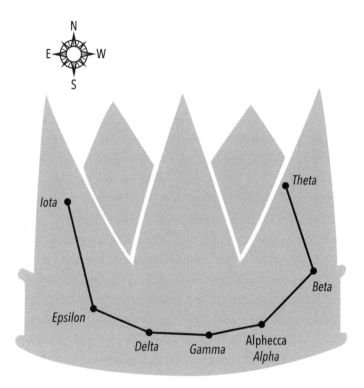

Figure 4.5. Corona Borealis is the celestial crown of authority, knowledge, and spirituality.

In Celtic Welsh mythology, Corona Borealis was called Caer Arianrhod, which means "the castle of Arianrhod." It is also known as the Silver Wheel (of birth, death, and rebirth) and associated with the moon. The semicircle of this constellation's stars is thought to have been associated with Arianrhod because its shape suggests a wheel as well as a crescent moon.

Ptolemy equated this constellation with Venus and Mercury. Corona Borealis is said to influence one's ability to assume a position of command.

Notable Star in Corona Borealis

Official Designation: Alpha Coronae Borealis

Traditional Names: Alphecca; Gemma

Pronunciations: al-FECK-ah; JEM-uh

Alphecca is a double star: Alpha-1 is white and Alpha-2 is yellow. Its traditional name comes from an Arabic root word that means "break" or "broken," which is a reference to the shape of the constellation as a broken circle. Corona Borealis was also regarded as a broken bowl or dish. This star's Latin name is Gemma, which means "jewel." Another Latin name that was used during the Middle Ages is *Gnosia Stella*, or just *Gnosia*, meaning "the star of knowledge." Alphecca was one of the fifteen important fixed stars in medieval magic. It is associated with artistic abilities and goodwill. The alpha star in Corona Australis the Southern Crown also has Alphecca as part of its name.

Magical Interpretations and Uses for Corona Borealis

Because Corona Borealis is associated with a crown, let's first consider this symbolism. A crown indicates a person's position and/or authority and his or her ability to provide protection. With this in mind, we can call on Corona Borealis for guidance and protection in our everyday endeavors. As a celestial crown, Corona Borealis can take on a spiritual meaning and aid in illuminating our spiritual paths. In addition, its name Gnosia comes from the Greek *gnonai*, "to know" or "to perceive," which enhances its role for providing guidance. Its association with knowledge and perception makes the energy of Corona Borealis apropos for divination and any form of psychic work.

A crown is also symbolic of success and the ability to rise above things that might otherwise hold us back. With this in mind, light a semicircle of seven candles as you sit in front of your altar. Imagine the stars of Corona Borealis shining in the heavens and then slowly descending to form a crown on your head as you visualize your success.

Similarly, instead of a crown you can choose to see Corona Borealis as a bowl or dish as you sit in front of your altar. Cup your hands, one in the other with palms upward, as you visualize the bowl of Corona Borealis pouring silvery light down into your hands. With this light comes blessings and the ability to attain what you desire.

By equating Corona Borealis with the Silver Wheel of Arianrhod, this constellation becomes instrumental in turning the Wheel of the Year as the world is reborn after winter. Corona Borealis can be used to draw down the energy of Arianrhod, for although she is associated with the moon, she is a goddess of night and magic. Like other spring constellations, Caer Arianrhod also carries the theme of duality and balance as seen in the goddess's sons Lleu (associated with light) and Dylan (associated with dark).

★ ★ ★

Corvus: The Crow/Messenger from Other Realms

Pronunciations: Corvus (CORE-vus); Corvi (CORE-vee)

Visible Latitudes: 60° North to 90° South

Constellation Abbreviation: Crv

Bordering Constellations: Crater, Hydra, Virgo

Description: The most recognizable pattern is an uneven rectangle or trapezoid of four stars with two spokes protruding at different angles.

To Find: Corvus is low in the sky near the horizon below Virgo. Locate the bright star Spica in Virgo, and as the saying goes "curve down to Corvus." It is just above the tail of Hydra the Water Snake.

This constellation is located in the Southern Hemisphere, but can be seen low in the sky in the north. Meaning "crow" or "raven" in Latin, *Corvus* is the genus name for both of these birds. The Babylonians knew this constellation as the Great Storm Bird and depicted it perched on a sea serpent, pecking at its tail. It was associated with Adad, a god of rain and storms, because the constellation was most visible just before the rainy season. The ancient Arabs originally regarded this constellation as a camel, but after European influence it became known as a crow or raven.

In Greek mythology, the crow was sacred to Apollo and Athena. Corvus shares a legend concerning Apollo with the constellations Crater the Cup and Hydra the Water Snake. Versions of the story vary as to whether Apollo was simply thirsty or needed water for an altar sacrifice. At any rate, he sent Corvus with a cup to get water. Distracted from his mission by a fig tree, the crow decided to wait until the fruit ripened. After his feast, Corvus finally fetched the water and returned to Apollo. To bolster his alibi for being late, Corvus took a snake along to show Apollo that he had to subdue it to get the water. Angry at the obvious ruse, Apollo threw the crow, the cup, and the snake into the sky. He also put a curse on Corvus by placing him near the cup, but never allowing him to reach it. Perpetual thirst is said to be the reason why crows and ravens make raspy sounds.

Another well-known story about a crow and a cup is one of Aesop's fables. It tells about a thirsty crow that could not reach the water in a cup so he carefully dropped stones into it to raise the water level so he could drink. This story illustrates crows as intelligent and patient rather than greedy and deceitful.

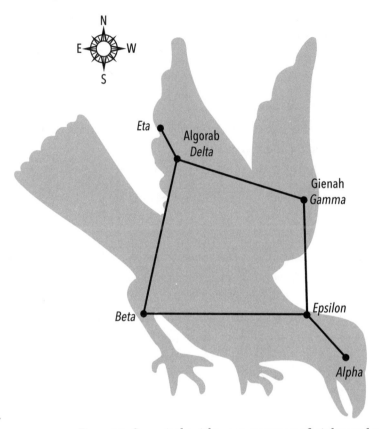

Figure 4.6. Corvus is the night crow, messenger of wisdom and prophecy.

According to Ptolemy, the energy of Corvus is equated with Mars and Saturn. Its magical influence was believed to give rise to craftiness, ingenuity, and prophecy.

Notable Stars in Corvus

Official Designation: Gamma Corvi

Traditional Name: Gienah

Pronunciation: JEEN-ah

Sharing its traditional name with Epsilon Cygni in the constellation Cygnus the Swan, Gienah is a blue-white star and the brightest in the constellation. Also spelled Giena, the name comes from an Arabic phrase meaning "the right wing of the crow." In depictions of the constellation, this star is on the right side; however, from the bird's perspective it is actually his left wing. Together, both wing stars, Gienah and Algorab, were considered one of the important fifteen fixed stars used in medieval magic.

Official Designation: Delta Corvi

Traditional Name: Algorab

Pronunciation: ALL-gor-ab

Algorab is a double star on the crow's other wing. Delta Corvi-1 is a yellow-white star and Delta Corvi-2 has been described as lilac or purple. This star's traditional name comes from Arabic and means "crow" or "raven." Along with Gienah, Algorab points toward Spica in Virgo and is one of the fifteen fixed stars.

Magical Interpretations and Uses for Corvus

To the Greeks, crows and ravens functioned as messengers of the gods, carrying wisdom and secrets. The crow was considered an attribute of Mithras and a mischievous associate of Loki and Mercury. As an important symbol of war and death in Celtic mythology, the triple goddess known as the Morrígan (Morrígan, Badb, and Macha) appeared as a crow over battlefields. Because crows are scavengers and feed on dead things, they came to be regarded as messengers from the otherworld.

The crow is considered to be on the edge between light and dark, life and death. In the cycle of life and death, death gives way to new life. Like so many of the spring constellations, Corvus has a dual nature. He is the solar crow, white by day (we cannot see him) and black by night (his stars reveal him). Perhaps it is the crow's duality that makes him mysterious.

The Celts believed crows and ravens to have powers of divination and prophecy. For this reason, we can call on Corvus to aid in divination and otherworld journeys. If you have seen crows in your neighborhood, go to that place on the day you plan your divination session. Wait, watch, and listen. Quite often more than one will appear. When you see or hear one, close your eyes and reach out to it with your energy. Don't try to do anything, just observe what you feel. Be patient, and when you sense a slight shift in energy begin to chant: *"Corvus, Corvus, black as coal; May your wisdom touch my soul. When you fly by dark of night; May my energy with you take flight."* When you are back home and ready for your divination session, take a moment to settle yourself and recall the feeling of that energy shift.

As an alternative to locating a crow, place a picture of one on your altar, and then take time to center and ground your energy. When you are ready, light a black candle and recite the Corvus chant noted above. Proceed with your divination or journeying.

If you sense that the crow is a totem animal for you, find a picture or figurine of a crow to keep on your altar. In the early spring, place it in a window exposed to the southwestern sky at night as you recite the Corvus chant. Draw down the energy of Corvus and release it into the picture or figurine. Leave it on the windowsill overnight, and then return it to your altar so your sacred space will absorb the energy of the constellation for a boost of crow magic.

Crater: The Cup/Chalice of the Goddess

Pronunciations: Crater (KRAY-ter); Crateris (kray-TER-iss)

Visible Latitudes: 65° North to 90° South

Constellation Abbreviation: Crt

Bordering Constellations: Corvus, Hydra, Leo, Virgo

Description: Six of this constellation's stars form a crescent shape. Two other stars suggest an uneven rectangle attached to the crescent.

To Find: Locate the bright star Spica in Virgo. Crater is to the southwest of Spica and west of Corvus the Crow. Crater and Corvus are tucked into a curve of Hydra the Water Snake.

The name Crater is Latin and means "cup" or "bowl." On old star maps this constellation is often depicted as a double-handled chalice. In some myths, it represented the cup or goblet of Apollo. See the entry for Corvus for the story about Apollo that links Crater with Corvus and Hydra. According to other legends, this constellation was called the Bowl of Bacchus. Due to its association with Dionysus, who is also called Bacchus, Crater was also linked with sensuality and enjoyment. Additionally, this constellation was known as Ganymede's Cup and represented the sacred goblet used to serve nectar to the Olympian gods.

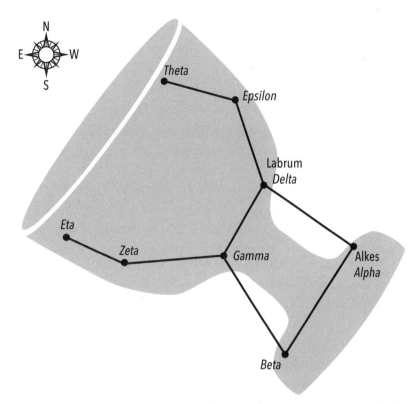

Figure 4.7. Crater represents the chalice of life and is a primary symbol of the Goddess.

To the Babylonians, this constellation represented the cup of Ishtar, who used it "for brewing fertility." [14] Noting that the Nile flooded at the time of year when Crater was visible, the Egyptians also associated it with fertility. To the Celts, this constellation represented the prophetic drinking vessel of Bran's cauldron.

Ptolemy equated the energy of Crater with Venus and to some degree with Mercury.

Notable Stars in Crater

Official Designation: Alpha Crateris

Traditional Name: Alkes

Pronunciation: AL-kez

Alkes is an orange star located on the base or stem of the cup. Its name comes from Arabic and means "the wine cup." Variations in the spelling of this name include Alker and Alhes. This star is associated with spirituality and mysticism.

Official Designation: Delta Crateris

Traditional Name: Labrum

Pronunciation: LAY-brum

This yellow star is actually the brightest in the constellation. Labrum is one of the six stars that form the bowl of the cup. In Latin its name means "lip." Labrum is associated with psychic abilities.

Magical Interpretations and Uses for Crater

Classic mythology links the three constellations of Corvus the Crow, Crater the Cup, and Hydra the Snake with Apollo. However, for twenty-first-century Pagans and Wiccans, all three are ancient and powerful symbols of the Great Mother Goddess.

The cup, or chalice, is the well-known magical and ritual tool for the element water. It is the symbol of the Goddess and of feminine powers. The cup represents the vessel of plenty, nourishing breast, and sacred womb. It is the chalice of life and an appropriate symbol for spring. Because the chalice on our altars represents the Goddess, we can use it to draw down the energy of this constellation to amplify her presence. This can also be used to bless your altar and ritual space.

Place a plain black cloth on your altar and a single candle in whatever color represents the Goddess to you. Use star-shaped glitter/confetti to lay out the star pattern of Crater. Fill your ritual chalice with water to represent the Goddess's sacred moisture. Stand with the cup in both hands at arm's length in front of you as you draw down the energy of Crater, and then say:

"May the abundance of your earth bring stability and strength to this altar.

May the freshness of your air bring clarity of mind to this altar.

May the heat of your fire bring healing and passion to this altar.
May the depth of your water bring forth your ancient wisdom.
Great Goddess of all things, bless this space and make it sacred in your eyes.
So mote it be."

Because the cup/chalice represents a womb, Crater is an appropriate constellation for spell-work relating to fertility. Also, as the bowl of Bacchus or the cup of Dionysus, call on Crater's energy at Beltane or for any magic relating to sensuality in the lusty month of May.

Hydra: The Water Snake/Symbol of the Goddess and Transformation
Pronunciations: Hydra (HIGH-druh); Hydrae (HIGH-dry)

Visible Latitudes: 54° North to 83° South

Constellation Abbreviation: Hya

Bordering Constellations: Cancer, Canis Minor, Centaurus, Corvus, Crater, Leo, Libra, Monoceros, Virgo

Description: Hydra winds its way down from the Northern into the Southern Hemisphere. Resembling a long, twisting snake, Hydra has three distinctive parts: the head, the back where Crater the Cup is located, and the tail where Corvus the Crow is located.

To Find: From the two stars in the bowl of the Big Dipper near the handle, draw an imaginary line southwest to the bright star Regulus at the bottom of the sickle shape in Leo. Continue that line to the next bright star, which is Alphard in Hydra. The compact group of stars to the northwest of Alphard marks the head of Hydra. The body of Hydra winds to the east and south under Crater and Corvus.

Hydra is the longest and largest constellation and is actually in the Southern Hemisphere. The name Hydra is Latin and means "water snake." In Greek mythology, Hydra appears in the story about Apollo's crow, which is represented by Corvus. The details of this story are included in the entry for the Corvus constellation.

Hydra is also associated with the serpent of Lake Lerna, the famed water snake with multiple heads. In different versions of the story, Hydra had anywhere from seven to one hundred heads and when one was cut off, two would grow in its place if the wound had not been seared by fire. In classic mythology, killing Hydra was one of the twelve labors, or feats, to be performed by Hercules. In addition to being the offspring of Echidna (the mother of all monsters), Hydra was the foster child of Hera, goddess of the sky and the heavens. Lake Lerna was believed to be a portal to the underworld, and Hydra served as its gatekeeper. Several of Hercules's tasks involved killing animals sacred to Hera, which we see today as attempts to destroy all symbols and traces of Goddess worship.

To the ancient Arabs and Hebrews, the snake represented all that was evil. This is reflected in their names for the constellation, which generally meant "the abhorred." Quite the opposite for the Egyptians to whom Hydra represented the life-giving River Nile. In medieval magic, Hydra was considered a bringer of wisdom and riches. Ptolemy equated Hydra with Saturn and Venus.

Notable Star in Hydra

Official Designation: Alpha Hydrae

Traditional Name: Alphard

Pronunciation: AL-fard

Alphard is a red star located on the forward part of the snake's body. Its name is Arabic and means "the solitary one." Alphard was also known as the Backbone of the Serpent, however, astronomer Tycho Brahe called it Cor Hydrae, "the Heart of Hydra." This star has been associated with untamed emotions and darker passions.

Magical Interpretations and Uses for Hydra

The Great Mother Goddess represents the principle of life, reproduction, and power. The snake and bird are two of the oldest and most important creatures associated with the Goddess. For our purposes here, we will only discuss the snake. The obvious cycles of shedding its skin linked the snake with the moon and fertility. Further, being associated with water connected snakes with the life-giving, primordial energy of the Goddess. Ironically, Hydra is best seen during the month of March when St. Patrick is honored for driving the snakes (Goddess worship) out of Ireland. Because of this, I like to represent the ancient power of the Goddess on my altar with a small plastic snake or two on March 17.

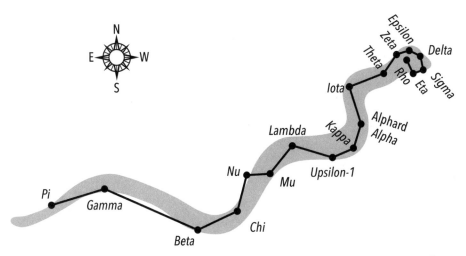

Figure 4.8. Hydra the snake represents one of the ancient and most powerful symbols of the Great Goddess.

The snake was also a symbol of healing, and it has been suggested that the name of the healer god Asclepius was derived from a Greek word for serpent. Hygeia, the goddess of health and the daughter or companion of Asclepius, was often depicted with a serpent drinking from a cup that she held for it. In addition, both of these deities were depicted with serpents winding around their staffs. A good way to draw the energy of Hydra into a healing circle is with a spiral dance. The coiling motion of this dance is reminiscent of a snake's movement. Begin by standing in a circle with everyone holding hands. The lead person will drop the hand of the person to her or his left and begin to walk to the left inside the circle, leading the others in a chain. The leader continues to spiral inward making the circles smaller and smaller. When the center is reached, the leader turns to the left, leading the chain outward between rows of other dancers who are still spiraling inward. When the dance is finished, have everyone join hands in a circle again to focus their minds and energy toward the person who needs healing.

Because snakes are symbolic of change, this constellation can help bring change into your life and help initiate transformation. Position a small plastic snake in a shallow bowl or saucer of water, which is an element symbolic of change. Place it outside or on a windowsill on a clear night as you ask for Hydra's help in raising energy for your intention.

Repeat the following three times: *"May the flow of Hydra's energy and the radiance of the Great Mother bring the changes I seek."* In the morning, pour the water on the ground or in a potted plant if you live in an apartment. Put the snake on your altar or another place where you will see it frequently and be reminded of your purpose.

Leo: The Lion / Solar Power and Protection

Pronunciations: Leo (LEE-oh); Leonis (lee-OH-niss)

Visible Latitudes: 90° North to 65° South

Constellation Abbreviation: Leo

Bordering Constellations: Cancer, Crater, Hydra, Ursa Major, Virgo

Description: Six stars form a backward question mark pattern also known as the Sickle of Leo. This pattern forms the lion's head and mane. Regulus, the brightest star in the constellation, appears as the period in the question mark. To the east of the question mark is a triangle that forms Leo's back leg and tuft of fur at the end of his tail.

To Find: Draw an imaginary line from the North Star at the end of the handle of the Little Dipper toward the two stars in the bowl of the Big Dipper, opposite the handle. Continue that line to the southwest to the bright star of Regulus located at the bottom of the question mark / sickle. The bright star to the east of Regulus is Denebola, which marks the end of Leo's tail.

Leo is one of the oldest recognized constellations. From approximately 6000 to 3000 BCE, it was prominent during the summer solstice, giving it, and lions in general, a connection with the sun. The solstice also coincided with the flooding of the Nile, which gave Leo the status as a protective deity of sacred water as well as fertility of the land. The lion-head motif often seen on fountains carries this symbolism of protecting water.

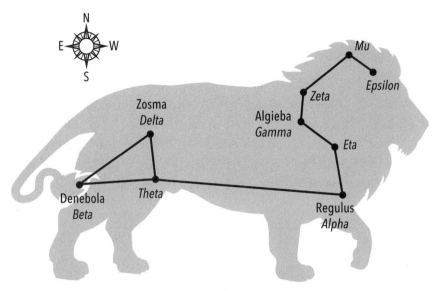

Figure 4.9. Leo is the harbinger of spring and summer.

The ancient Persians and Mesopotamians regarded this constellation as a lion, and in mythology Leo represented the lion that Hercules killed as one of his twelve tasks. As the story goes, this lion lived near the town of Nemea southwest of Corinth in Greece. Weapons could not pierce its skin, but Hercules was able to strangle the animal. After skinning it, he wore the pelt as a protective cloak. Because the lion was the king of beasts, his spirit was placed among the stars.

According to Nicholas Culpeper, Leo influences the heart, back, spine, and thymus gland. Ptolemy equated the energy of most of the stars in Leo with Saturn and Mars, Venus, or Mercury, depending on their location. According to modern astrology, Leo is equated with the energy of the sun.

Notable Stars in Leo

Official Designation: Alpha Leonis

Traditional Name: Regulus

Pronunciation: REG-you-luss

Regulus is the brightest star in Leo and marks the lion's heart. It is actually a triple star. Alpha-1 is blue-white, Alpha-2 is orange, and Alpha-3 is red. In Latin, *Regulus* means "little king" or "little prince." This star's Arabic name means "the heart of the lion." Regulus was one of the royal stars of Persia and was called the Guardian of the North. In times past, navigators used it to determine longitude. Regulus was also one of the fifteen important fixed stars in medieval magic and associated with power and success.

Official Designation: Beta Leonis

Traditional Name: Denebola

Pronunciation: deh-NEB-oh-lah

With a name from Arabic meaning "the lion's tail," Denebola marks the tuft of fur on Leo's tail. This blue star is the western anchor point for the Spring Triangle and Great Diamond asterisms. Denebola is associated with nonconformity that stimulates creativity and ingenuity.

Official Designation: Gamma Leonis

Traditional Name: Algieba

Pronunciation: al-JEE-bah

Algieba is a double star located on the lion's neck, despite its name meaning "the forehead." In Latin it was called Juba, which has a more appropriate meaning of "mane." Gamma-1 is orange and Gamma-2 is yellow. It is associated with promoting love and providing protection against enemies.

Official Designation: Delta Leonis

Traditional Names: Zosma; Duhr

Pronunciations: ZOSS-mah; door

Zosma is a blue-white star that marks the lion's hindquarters. Its traditional name is Greek and means "the girdle." Its other name, Duhr, comes from an Arabic phrase that means "the lion's back."

Magical Interpretations and Uses for Leo

Leo is a harbinger of spring, the increasing light, and the promise of warmer days to come. If you have a fountain or water feature in your home or garden, ask Leo to be its guardian and place an image or figurine of a lion nearby. The star pattern of Leo drawn on a fence or other fixture in a vegetable garden invites fertility and abundance. As a Babylonian royal star and Guardian of the North, call on Regulus when evoking that direction in spring rituals.

Although most often equated with sun gods such as Amun and Mithras, the lion is also symbolic of feminine powers—lionesses are the primary hunters. Some historians believe that the Babylonians may have regarded this constellation as a lioness. In addition, lions were depicted at the feet of Inanna and Ishtar. Call down the energy of Leo to help you connect with both male and female deities.

More than anything, Leo represents power, protection, and courage—attributes that Hercules wanted to tap into after slaying the lion of Nemea by wearing its pelt as a cloak. Call on Leo for help in building energy for spells of protection and to foster courage. To do this, you will need a piece of paper, a cotton swab, a bowl, a sharp knife or athame, and a lemon. The color and shape of the lemon is symbolic of the sun. Cut the lemon in half and squeeze the juice into a bowl. Dip the swab into the juice and use it to write what you seek on the paper. When the juice dries, the writing will become invisible. Fold the paper and place it on your altar until the next starry night. For the second part of this you will need a candle and a cauldron or a safe place to burn the paper. Go outside or sit by a window where you can see the stars. Hold the paper between your palms and say: *"Powerful Leo, King of all; Know my wish, hear my call. As you stride across the heavens in regal glory; Your blessings I ask, to fulfill this story."*

Repeat this two more times as you drawn down the energy of Leo. Continue to hold the paper in your hands as you visualize what you seek. When it is clear in your mind, go to your altar and light a candle. Hold the paper close enough to warm it, but not burn it. The writing will become visible again. Once it is, allow the paper to catch fire and then drop it into your cauldron.

Libra: The Scales/Balance and Justice
Pronunciations: Libra (LEE-brah); Librae (LEE-bray)

Visible Latitudes: 65° North to 90° South

Constellation Abbreviation: Lib

Bordering Constellations: Hydra, Lupus, Ophiuchus, Scorpius, Serpens, Virgo

Description: The most notable shape in Libra is a large trapezoid of four stars.

To Find: Locate Spica in Virgo at the bottom of the Great Diamond asterism and draw an imaginary line to the east. Libra is a relatively faint constellation east and slightly south of Virgo.

Libra is a constellation of the Southern Hemisphere that can be seen throughout most of the north. Originally the Greeks considered Libra part of Virgo and associated the whole constellation with Themis, the goddess of justice. Later they considered Libra part of Scorpio and named it Chelae Scorpionis, "the claws of the Scorpion." The Romans regarded Libra as a separate constellation that represented the scales of justice. The name Libra is Latin and means "the weighing scales."

To the Egyptians, the constellation represented the scales on which one's heart was weighed after death. A heart weighing more than a feather indicated that the deceased had done some bad deeds and would be destined to wander in oblivion rather than enjoy a good afterlife. Libra also represented the balance of the sun in the upper- and under-worlds. In China and India, the constellation represented balance, and the Sumerians called it the Balance of Heaven.

According to Culpeper, Libra influences the kidneys, lumbar region, skin, adrenal glands, and back of the body. According to modern astrology, Libra is equated with Venus.

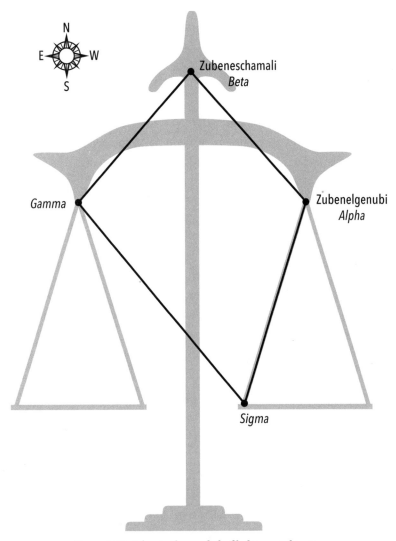

Figure 4.10. Libra is the symbol of balance and justice.

Notable Stars in Libra

Official Designation: Alpha Librae

Traditional Name: Zubenelgenubi

Pronunciation: zoo-BEN-el-jeh-NEW-bee

Zubenelgenubi is a double star in which Alpha-1 is yellow and Alpha-2 is blue. Its name means "the southern claw" in Arabic, which is a holdover from the time when it was considered part of Scorpius. Now it usually marks a point on the balance arm in depictions of Libra. In the past, this star was believed to have negative influences.

Official Designation: Beta Librae

Traditional Name: Zubeneschamali

Pronunciation: zoo-BEN-esh-ah-MAHL-ee

Zubeneschamali is the most visible star in the constellation. It is a blue star that usually appears somewhat greenish in color. Its name means "the northern claw" in Arabic. It usually marks the top of the scales in depictions of Libra. Zubeneschamali is associated with honor and ambition.

Together these two stars represent light and dark, the quintessential yin and yang of balance.

Magical Interpretations and Uses for Libra

Although the spring equinox may occur before Libra is visible to those of us in the Northern Hemisphere, the constellation is a reminder that this is a time to bring our lives into balance. I like the Egyptian idea of weighing one's heart. Not to determine if we should walk in oblivion, but as an introspective exercise to bring our thoughts and emotions into balance. Draw the configuration of the Libra constellation at the top of a sheet of paper. On one side list your negative thoughts or emotions, and on the other side list things for which you are grateful. Do this quickly without giving it a lot of thought. Which side has more entries? More in the gratitude column is okay, but if you have listed more negative things take time to figure out what is going on in your life and how you might make things better. Call on this constellation for help in coming into balance. For Ostara, create the Libra star pattern with glitter/confetti or candles on your ritual altar to symbolize balance.

For aid in bringing your health into balance, draw the constellation on the outside of a mug that you use for herbal tea or other healthy drink. As you pour the beverage into the mug, say: *"Stars of balance, I call on thee; Bring strength and good health to me."*

To balance the energy of your living space, lay out four pieces of amethyst, citrine, clear quartz, or turquoise (or one of each) in the trapezoid pattern of the constellation. These gemstones are associated with balance. You can also use smoky quartz, jade, aquamarine, or blue tourmaline, as these are associated with Libra.

Just as ancient people equated Libra with goddesses of justice, call on the power of this constellation for support in legal matters or whenever you need to right a wrong. Light a green, pink, or white candle that was prepared with chamomile, geranium, patchouli, or spikenard oil. Sit in front of your altar and as you light the candle, whisper: *"Libra, Libra, shining in the night."* As you gaze into the flame, review your situation and then visualize the outcome you want to occur. When you are finished, put out the candle as you say: *"Libra, Libra, shining in the night; Stars of justice, help make things right."*

Ursa Major and Ursa Minor: The Great Bear and the Little Bear/
  Turning the Wheel of the Year
Pronunciations: Ursa Major (ER-suh MAY-jor); Ursae Majoris (ER-sigh mah-JOR-iss)

Visible Latitudes: 90° North to 30° South

Constellation Abbreviation: UMa

Bordering Constellations: Boötes, Canes Venatici, Draco, Leo

Pronunciations: Ursa Minor (ER-suh MY-nor); Ursae Minoris (ER-sigh my-NOR-iss)

Visible Latitudes: 90° North to 10° South

Constellation Abbreviation: UMi

Bordering Constellations: Cepheus, Draco

Descriptions: The stars of the Big Dipper are a noticeable pattern within Ursa Major and form the hindquarters and tail of the Great Bear. A triangle of stars represents his head and snout. The four stars or bowl of the Little Dipper represent the body of the Little Bear, and the three stars in the handle its overly long tail.

To Find: The dipper asterisms are the most familiar shapes in the northern sky. Even though their positions change with the seasons, they are easy to find.

Around the world and throughout time, Ursa Major was most often regarded as a bear. The ancient Phoenicians and Persians regarded Ursa Major as a bear, while the Egyptians related it to what they knew and called it the Hippopotamus as well as the Dog of Set. The Algonquian people of North America called the constellation the Bear and the Hunters. However, because bears do not have long tails they considered the three stars of the Big Dipper's handle as hunters following the bear. In Zuni lore, this constellation was called the Great White Bear of the Seven Stars.

Since ancient times, both constellations have been used for navigation to determine the direction north. Greek author Homer wrote that the hero Odysseus had to keep the bear on his left in order to sail east. The Celts also used the celestial bears for navigation.[15]

The Big Dipper is the name most widely used in the United States for the asterism in the tail and hindquarters of the Great Bear. During and after the Civil War, it was known as the Drinking Gourd and used by slaves to find their way to safety in the north. In Europe, the Big Dipper is known as the Plough, "plow." In Old English, it was called Carles's Waen or Charles's Wain, "wagon." The Welsh refer to it as Cerbyd Arthur, "Arthur's Wain." The four stars that form the bowl of the dipper are considered as the wheels of the wagon and the three stars of the handle the horses or oxen pulling it. Instead of a plow, in Latin it was called *Septentriones*, "the seven oxen." To the Babylonians, the Big Dipper was the Wagon of Heaven. In addition, because the constellation did not rise or set it was associated with eternity and the afterlife.

To the Greeks and Romans, Ursa Major represented Callisto, the handmaiden of Artemis (Greek myth), and a member of the Dianic cult of women (Roman myth). In one version of the Roman legend, Jupiter's affection for Callisto made his wife Juno jealous. To save her from Juno's wrath, he turned Callisto into a bear to disguise her. Hearing about this, Juno tricked Diana into hunting the bear. Once more Jupiter intervened and placed Callisto in the heavens. Callisto's son, Arcas, became the Little Bear of Ursa Minor. Of course, in the Greek legend Jupiter and Juno are Zeus and Hera. Another version of the Roman legend says that Diana turned Callisto into a bear and banished her because she broke the vow of chastity.

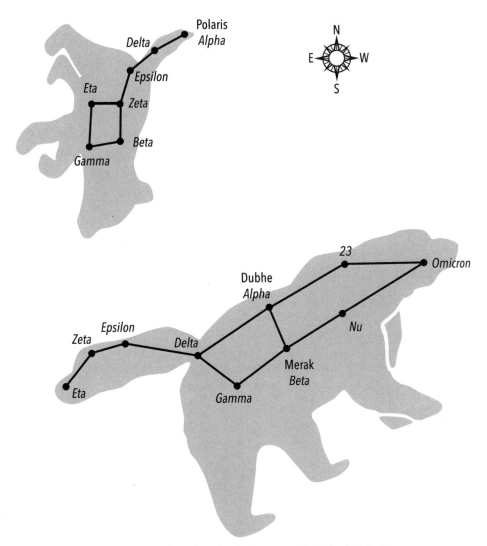

Figure 4.11. Circling the pole, the bears turn the Wheel of the Year.

The ancient Hindus associated the brightest stars of Ursa Major with the Seven Rishis and called the constellation *Saptarshi*, "the great sages."

Ursa Minor was regarded as a bear only after it was considered a constellation in its own right during the sixth century BCE. Prior to that it was regarded as a wing on the back of Draco the Dragon. It is thought by some historians that at one time Ursa Minor may have been called the Hesperides in honor of the nymphs who tended the garden of paradise and Hera's tree that bore golden apples. This name fits with legends of Hercules that play out amongst the stars because one of his tasks was to slay the dragon and steal the apples.

According to Ptolemy, the energy of Ursa Major is equated with Mars, while Ursa Minor is equated with Saturn and to some degree Venus.

Notable Stars in Ursa Major

Official Designation: Alpha Ursae Majoris

Traditional Name: Dubhe

Pronunciation: DOOB-huh

Official Designation: Beta Ursae Majoris

Traditional Name: Merak

Pronunciation: MER-ak

Dubhe, an orange star, and Merak, a blue-white star, are known together as the Pointers. They are located opposite the handle and mark the outer end of the bowl of the Big Dipper. They are called the Pointers because they point toward Polaris in Ursa Minor in one direction and toward Regulus in Leo in the other. Located on the back of the Great Bear, the name Dubhe is Arabic and means "bear." Although it is not the brightest star, its position as the first of the seven stars in the Big Dipper made it worthy of the alpha designation. The name Merak also comes from Arabic and means "the loin." It is located on the bear's flank. The Pointers have also been known as the Keepers.

Notable Star in Ursa Minor

Official Designation: Alpha Ursae Minoris

Traditional Name: Polaris

Pronunciation: poe-LAHR-iss

Commonly known as the North Star and the Pole Star, Polaris is at the tip of the Little Bear's tail. It is actually a double star, with one bright, yellow-white star and a smaller, dimmer companion of the same color. This is the nearest star to the celestial pole and has been used for navigation for centuries. Polaris has been known by many names including: Lodestar, the Steering Star, Stella Maris (sea star), and the Gate of Heaven.[16] It was one of the fifteen fixed stars of medieval magic and used for protection against spells.

Magical Interpretations and Uses for Ursa Major and Ursa Minor

Circling the pole, the bears turn the Wheel of the Year. Across time and cultures bears have represented rebirth, renewal, power, and healing. Their hibernation was regarded as having power over life and death. Hibernation comes to an end as mother bears leave their dens with cubs that were born in the dark winter months. This young life emerging from under the ground is symbolic of reawakening and renewal. In the north, receding sea ice brings polar bears to land. It is no surprise that to indigenous people of Siberia and Alaska the bear was associated with elemental forces and the cycles of nature.

We can call on the celestial bears for energy to stir the elemental forces that hasten spring. Just as the she-bear represents a healing aspect of the Great Goddess, we can call on Ursa Major and Ursa Minor to bring healing to land that may have been damaged by the harshness of winter. We can also ask the she-bear to protect young animals, especially those born in the early spring when the weather is so unpredictable.

With the other interpretation of the stars representing a wagon or plow, we look to Ursa Major as a reminder of our task in making the fields and gardens ready for planting. Combined with the constellation Boötes, this plow and plowman stand as reminders of our ancestors whose livelihoods and survival were intimately connected with the earth. In addition, we can ask that their combined energies add blessings to the land and lead to a bountiful harvest in the autumn.

Now or at any time of year when travel is on your agenda, whisper a prayer to Polaris before departing. *"North Star, Pole Star, shining bright; High above the world at night. 'Ere I travel by land, air, or sea; Stella Maris guide and protect me."*

Virgo: The Virgin/Maiden and Mother Nurture the World

Pronunciations: Virgo (VER-go); Virginis (VER-jin-iss)

Visible Latitudes: 80° North to 80° South

Constellation Abbreviation: Vir

Bordering Constellations: Boötes, Corvus, Crater, Hydra, Leo, Libra, Serpens

Description: Positioned between Leo and Libra, this constellation forms a stick figure.

To Find: Follow the handle of the Big Dipper and according to the saying, "arc to Arcturus and then speed on to Spica." Spica is the brightest star in Virgo and one of the brightest in the sky.

Although Virgo is usually called the Virgin, at one time the word also referred to "any virtuous matron." [17] To the Greeks and Romans this constellation represented Demeter and Ceres, respectively. Just as Virgo was becoming visible, an annual festival in honor of Ceres was held in Rome during the second week of April. To the Greeks, Virgo represented Persephone as well as Demeter holding a sheaf of corn/grain. In Europe, the word *corn* is a general term for grain and not specifically maize as it is in the United States. The Greeks also associated Virgo with Aphrodite and Rhea.

Another interpretation of this constellation links it to Astraea, the Greek goddess of justice. In Egypt, the constellation was depicted on the temples of Dendera and Thebes holding what some scholars have interpreted as a distaff, a tool used in spinning, instead of a sheaf of grain. Additionally, Virgo was associated with Isis and Hathor.

In the region of the Euphrates River, Virgo represented Ishtar and Astarte, while to the people of India the constellation was known simply as the Maiden. Virgo was also associated with Atargatis the Syrian goddess of fertility. In Peru, Virgo was known as the Earth Mother.

Virgo carries out her role as a goddess of grain since she appears in the night sky during planting season. It becomes a background constellation for the sun in September and October during the time of the autumn harvest.

Ptolemy equated most of the energy of Virgo with Mercury and Mars, and with Venus influencing her mid-section. In modern astrology, Virgo is equated with Mercury. According to Culpeper, Virgo influences the intestines, abdomen, solar plexus, and parasympathetic nervous system.

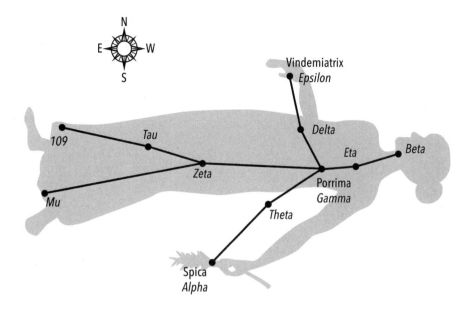

*Figure 4.12. Virgo represents the maiden and the mother,
the fertile earth, and the cycles of nature.*

Notable Stars in Virgo

Official Designation: Alpha Virginis

Traditional Name: Spica

Pronunciation: SPY-kah

Spica is a blue-white spectroscopic binary star. From Latin meaning "the head of grain," Spica represents a sheaf of grain in Virgo's left hand. Like Polaris, this star was used for navigation in ancient times. Ptolemy's geographical reference on the earth for Spica was the Fortunate Islands, present-day Canary Islands, which may account for the star's designation as the Fortunate One.[18] Spica was one of the fifteen stars important in medieval magic, representing abundance and protection against danger. This star is the southern anchor point for the Spring Triangle and the Great Diamond, which is also called the Great Diamond of Virgo.

Official Designation: Gamma Virginis

Traditional Name: Porrima

Pronunciation: pour-EE-mah

Porrima is a double yellow-white star named for a Roman goddess of prophecy.

Official Designation: Epsilon Virginis

Traditional Name: Vindemiatrix

Pronunciation: vin-duh-mee-AH-tricks

Vindemiatrix is the Latin translation of an older name that meant "grape gatherer." Astronomers believe that this star was named at a time when it may have risen with the sun during the grape harvest. It was also referred to as the Star of Bacchus. This yellow star is located on Virgo's right hand or arm. Today, astronomers use Vindemiatrix as an aid in finding the Virgo Cluster, a group of several hundred galaxies many of which are spirals.

Magical Interpretations and Uses for Virgo

With the return of spring, the process of renewal begins under the watchful gaze of Virgo, who symbolizes the womb and bounty of the earth as well as the (re)birth of spirit. As Demeter, she personifies the mother aspect of the Great Goddess, who rules the cycle of seasons and brings the gift of grain.

Virgo's association with Demeter and Persephone makes this constellation a natural fit for Ostara rituals when their story is told. Use the following to call in the energy of this constellation: *"Virgo, we call on you to provide divine sheaves of wheat to feed the hungry departed souls in the underworld, and with your power guide Persephone forth. May the daughter of Demeter step into the sunlight and rejoice in its warmth. And may she enjoy the light from your stars at night."*

Also as Demeter/Ceres, Virgo is goddess of fertile soil and can be called upon to aid our gardens. Place an image of Virgo among your plants or hang a pendant of the zodiac symbol in a tree or bush. You may find it more effective to create your own. Find an attractive rock in your garden, clean it up, and paint the Virgo star pattern on it. Place it in your garden at night when the constellation is overhead as you say: *"In my garden this token I place; I call on Virgo to bless this space. Help these plants flourish and grow; With your love, make it so."*

In Egypt, wheat was a symbol of Osiris. Coupled with the ancient connection with Isis, Virgo symbolizes the fundamental power of nature so that all that dies will be reborn. In addition to themes of regeneration and cycles, the story of Isis and Osiris is that of determination and love. Under the light of Virgo on Beltane night, call on the strength of their divine love to enrich your sabbat celebration.

If you use wine in your rituals, call on the Star of Bacchus, Vindemiatrix, to pour forth special blessings. Additionally, write this star's name on a fence or a rock in your garden for blessings to help the plants grow. Likewise, write Spica or Alpha Virginis on a small piece of paper to keep in your wallet or purse to bring prosperity.

On the night of a new moon when the stars are bright, set your divination tools on a windowsill and ask for the light of Porrima to bless them.

Chapter Five

THE SUMMER QUARTER OF JUNE, JULY, AUGUST

In June, we enjoy the longest days of the year just before the summer solstice. Even though the sun begins its journey south, the shortening length of daylight is imperceptible. While the world is lush and green, tune in to the rhythms of the natural world and become aware of the energy that surrounds you. Before the heat of high summer arrives, there is a brief period in which to enjoy the softness of this season and the magic of gentle starlight on warm summer nights.

Summer marks the time when the Great Goddess is in her full mother aspect as the fields and orchards ripen in July. Shimmering waves of heat rise skyward as we bask in the warmth of this season. This is the time to enjoy the freedom of lightweight clothing and the night chorus of crickets as you gaze skyward to work with the magic of the stars.

During August, humid weather blankets the land and imposes a slower pace to complete the annual cycle of growth. Roses may be fading, but lavender and chamomile are in their glory. Lazy days are offset by dazzling thunderstorms that tear the night sky asunder. Even as summer winds down, autumn is a distant horizon.

This time of year the night sky is called the Vega Quadrant because the star Vega in the constellation of Lyra is the brightest in the sky. An eagle, swan, scorpion, centaur, a mythical hero, a dragon, a musical instrument, a man with a snake, a sea-goat, an altar, and a wolf populate the summer canopy of constellations. Let's see what energy we can draw and what magic we can work with these constellations.

Some of the bordering constellations noted throughout this chapter may fall within other seasons. The directions given to locate constellations and stars assumes that the reader is facing south. Also, have a look at chapter 8 as some of the southern constellations listed only in that chapter may be visible to you.

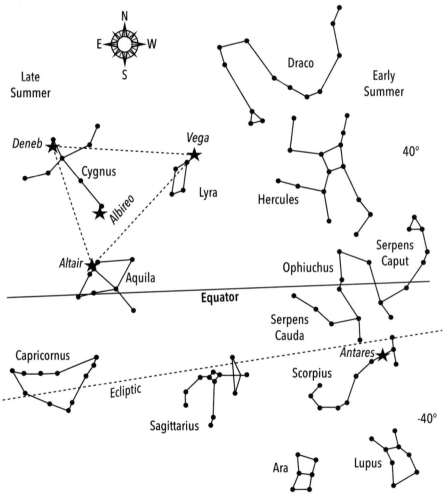

Figure 5.1. The summer sky. The dotted shape is the Summer Triangle asterism. The five brightest stars are also noted.

THE SUMMER CONSTELLATIONS

Aquila: The Eagle/Power of the Sun

Pronunciations: Aquila (ACK-ill-uh); Aquilae (ACK-ill-eye)

Visible Latitudes: 90° North to 75° South

Constellation Abbreviation: Aql

Bordering Constellations: Aquarius, Capricornus, Delphinus, Hercules, Ophiuchus, Sagittarius, Serpens

Description: A prominent line of three stars marks the upper part of the eagle's back. Extending to the southeast and northwest, other stars mark the tips of the wings. Stars for the tail extend toward the southwest.

To Find: Locate Vega, the brightest star in the summer sky. Draw an imaginary line to the southeast to the next bright star. This is Altair in Aquila.

Latin for "eagle," Aquila represents this majestic bird, and it has been recognized as such for over 4,000 years. Aquila is most often associated with the eagle companion and servant of Zeus who carried the god's thunderbolts. In one myth, Aquila lifted the young Ganymede to Olympus to be the cupbearer of the gods. In another legend, this constellation represented Aphrodite disguised as an eagle, which was part of a ruse. As an eagle, she chased Zeus, who was in the form of a swan, into the protective arms of Leda. Zeus later placed the image of an eagle and a swan among the stars to commemorate the event of his sexual conquest of Leda.

In quite a different story, Zeus sent an eagle to rescue the lyre of Orpheus after it was thrown into a river when the musician was murdered. He honored both bird and instrument by placing them in the sky. The lyre is represented by the Lyra constellation.

The alpha, beta, and gamma stars of this constellation have been called the Family of Aquila. The Persians called them the Striking Falcon, and in India this line of stars represented the footprints of the god Vishnu. Also, in an earlier Vedic myth of India an eagle messenger from the god Indra brought soma, a ritual drink of deities, to earth for humans. The Turks called the whole constellation the Hunting Eagle, and the Arabs called it the Flying Eagle.

Ptolemy equated the energy of Aquila with Mars and Jupiter, and noted that the constellation bestows clairvoyant abilities. Later astrologers equated Aquila with Saturn and Mercury as well as Uranus and Mercury. In medieval medicine, Aquila was credited with conferring honors and preserving traditions.

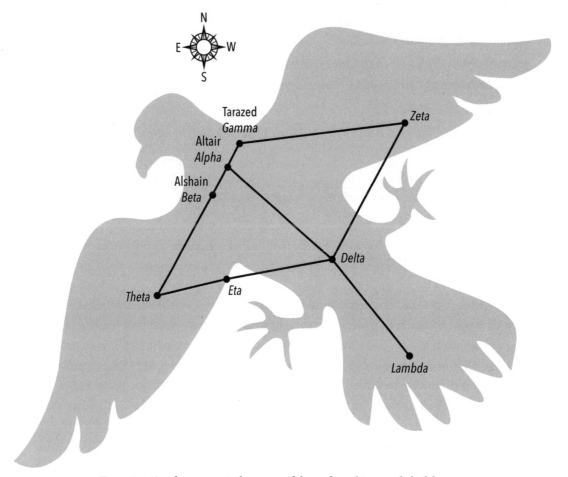

Figure 5.2. Aquila represents the power of the eagle and is a symbol of the sun.

Notable Stars in Aquila

Official Designation: Alpha Aquilae

Traditional Name: Altair

Pronunciation: AL-tair

Altair is the middle of three stars across the eagle's upper back. Its name is derived from the Arabic name of the constellation, the Flying Eagle. The Babylonians and Sumerians called Altair the Eagle Star, and Ptolemy's name for it was Aetus, "eagle." [19] In India, it is associated with the Hindu god Vishnu and Garuda, the eaglelike bird that he rides. Altair has also been called Vishnu's Star. This yellow star is the southern anchor point for the Summer Triangle asterism.

Official Designation: Beta Aquilae

Traditional Name: Alshain

Pronunciation: al-SHANE

Alshain is a double star composed of a yellow primary component with a red companion. Its position as one of the three distinctive stars across the eagle's upper back gives Alshain prominence. However, astronomers are baffled as to why it was given the beta designation because Tarazed, another of the three stars, is actually the second-brightest star in the constellation.

Official Designation: Gamma Aquilae

Traditional Name: Tarazed

Pronunciation: TAHR-ah-zed

As already mentioned, this orange star is the second brightest in Aquila and one of the three prominent stars. The names of Alshain and Tarazed come from a Persian phrase that was derived from Arabic and means "the balance." Tarazed and Alshain are on either side of Altair.

Magical Interpretations and Uses for Aquila

As a symbol of the sun, the eagle embodies the spirit of summer. It is one of the most sacred of animals to Native Americans. In Celtic lore, the eagle is one of the oldest and wisest creatures, and it is associated with prophecy and power. Celtic names for the eagle included *Suil-na Graine*, "eye of the sun," and *Lolair*, "guide to the air." [20]

As it is a bird of the sun, fire, and air, call on this celestial eagle to be a special presence at your Litha ritual. Lay out the pattern of Aquila's stars with glitter/confetti on your altar. Or, simply place three tea light candles in a row to represent the stars Altair, Alshain, and Tarazed. To call the eagle's energy say: *"Aquila, Aquila, Eye of the Sun; Today begins summer, your time has come. Bright as our day star, when you take flight; We ask for your presence, please join us this night."*

This invocation can be used when honoring Zeus, Jupiter, Mithras, Vishnu, Indra, Odin, or other powerful gods associated with the eagle. It is also an aid for getting in touch with the power of the terrestrial bird. Aquila's energy can be called upon for spells of success or whenever you need support in reaching new heights in your life.

As an eagle, Aquila is an aid for divination or any psychic work through which you seek wisdom. He is also a bridge to other realms and can act as a guide and a helper for astral travel. You can soar with a celestial eagle on the astral plane. In preparation the night before, draw the configuration of Aquila's stars on a piece of paper and then wrap the paper around or place it on top of your divination tools. Position them on your altar and say: *"Celestial eagle soaring high, through the dark of summer sky. These things upon my altar I place, for your power and wisdom to grace."* On the following night, create a protective circle. Burn frankincense, lavender, or sandalwood incense or light a black or indigo candle. Lay out the configuration of Aquila's stars on your altar with gemstones such as moonstone, obsidian, or clear quartz. As you unwrap your divination tools say: *"Eye of the sun, I bid you to come. With your guidance, this work will be done."*

Ara: The Altar/Heavenly Sacred Space

Pronunciations: Ara (AIR-uh); Arae (AIR-eye)

Visible Latitudes: 25° North to 90° South

Constellation Abbreviation: Ara

Bordering Constellations: Corona Australis, Lupus, Sagittarius, Scorpius

Description: Two adjoining and uneven rectangles represent the base of the altar and smoke rising from its top.

To Find: Locate the Summer Triangle and the star Altair at its southern anchor point. Draw an imaginary line toward the southwest through the Sagittarius constellation to Ara, which is located below the scorpion's tail.

This constellation's name is derived from Latin and means "altar"; however, in earlier days the word "ara" had two meanings. It could be an altar for hearth and home or a refuge for protection.[21] Originating with the Babylonians or possibly the Sumerians, this constellation was referred to in classic literature both as an altar and as a type of censer or incense altar that was common in the Middle East. Because the Romans considered Ara the altar on which Chiron the centaur made sacrifice, they called these stars Ara Centauri.

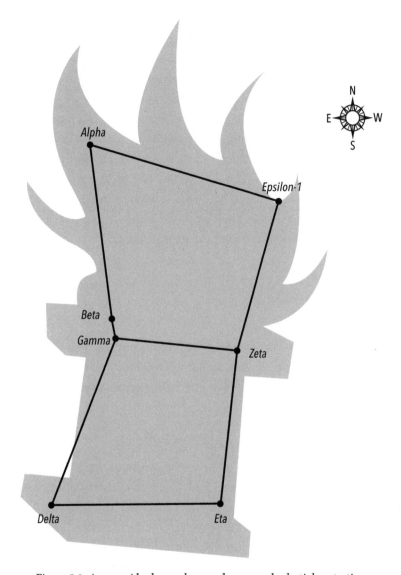

Figure 5.3. Ara provides heavenly sacred space and celestial protection.

    The Greeks considered Ara to be the altar upon which the Olympian gods made a vow of allegiance before their battle with the Titans. Vows performed at an altar were understood to be particularly binding, hence the custom today of taking marriage vows in front of an altar. Zeus placed the altar of the gods among the stars to commemorate their victory.

According to another Greek legend, Ara represented the altar that King Lycaon of Arcadia had dedicated to Zeus. Lycaon was known for his cruelties, and after he sacrificed a child on this altar Zeus turned him into a wolf. Another legend said that whenever a human sacrifice was made on Lycaon's altar a man was turned into a wolf or werewolf. On a lighter note, Ara was also regarded as the altar of Dionysus.

The Greeks imagined the ribbon of the Milky Way above Ara as smoke from the burnt offerings on the celestial altar. Astronomer Johann Bayer depicted an inverted version of Ara in his 1603 atlas, and for a time many star maps followed his lead.

Ptolemy equated this constellation with Venus and Mercury, and noted that it was associated with devotion. In medieval medicine it was regarded as an aid for maintaining chastity. In addition, the Greeks associated Ara with weather prediction, most likely because it became visible each year before the stormy season.

Notable Stars in Ara

Official Designation: Alpha Arae

Official Designation: Beta Arae

Although some of the stars were given names in other cultures, they did not carry over to the West. The blue alpha star is actually the second brightest in Ara; the orange beta is the brightest. The Chinese regarded these two stars as an asterism that they called Chu, which means "pestle."

Magical Interpretations and Uses for Ara

Although Ara is not visible to many of us in the Northern Hemisphere, it serves as a reminder of the importance of our altars. The concept of an altar dates to a time in prehistory when people began making offerings to their deities and needed a special place in which to do it. Altars have served as places where offerings and sacrifices are made—physically and symbolically. However, an altar is not just a thing that holds a collection of objects. Intention and energy transforms an altar into a space that transcends the mundane world. When we use an altar, we step outside the boundaries of our everyday lives. When we sit in front of an altar, we place ourselves in the presence of spirits, ancestors, and the Divine. We open ourselves to seek and search for answers to questions that guide our souls.

Summer is a good time to give thanks for warm weather and the growing season. A flat rock can serve as a simple outdoor altar. Draw Ara's star pattern on it and consecrate it as a sacred area of your garden by saying: *"Ara, shine your light on this space; Illuminate my sacred place."*

Because the meaning of the word *ara* could designate an altar for hearth and home, use your outdoor Ara altar to honor Vesta, Hestia, and Brigid, goddesses of the hearth who are also keepers of sacred flames. Place sprigs of lavender, blackberries, or dandelion flowers on it. The gemstones azurite, chrysoprase, or peridot are also appropriate tokens.

With the other Latin meaning of *ara* being a place of refuge and protection, Ara's energy can be used as a protective talisman. Draw the star pattern on ordinary stones or flowerpots that can be placed around your property or in any area where you feel protection is needed. If discretion is important, the star pattern can be drawn on the interior of a flowerpot before filling it with soil and plants.

Capricornus: The Sea Goat / Horned One of Abundance

Pronunciations: Capricornus (kap-rih-KORN-us); Capricorni (kap-rih-KORN-ee)

Visible Latitudes: 60° North to 90° South

Constellation Abbreviation: Cap

Bordering Constellations: Aquarius, Aquila, Piscis Austrinus, Sagittarius

Description: The star pattern of Capricornus looks like a fat letter V.

To Find: Start at Vega, the brightest star in the summer sky, and draw an imaginary line southeast to the next bright star, which is Altair in Aquila. Continue that line to the southeast and the next constellation is Capricornus.

Capricornus is located in a region of the sky that the ancients called the Sea. The constellation's name comes from the Greek *caper*, "goat," and *cornu*, "horn." Even though Capricornus is a rather faint constellation, its observation dates to the Babylonians, who referred to it as a goatfish. The Arabs and Persians regarded the constellation simply as a goat. The Romans associated it with the goddesses Venus and Vesta. They depicted Venus riding a sea-goat, and considered Capricornus to be under the protection of Vesta. In addition, images of sea-goats were incorporated into frescoes in the Roman catacombs.

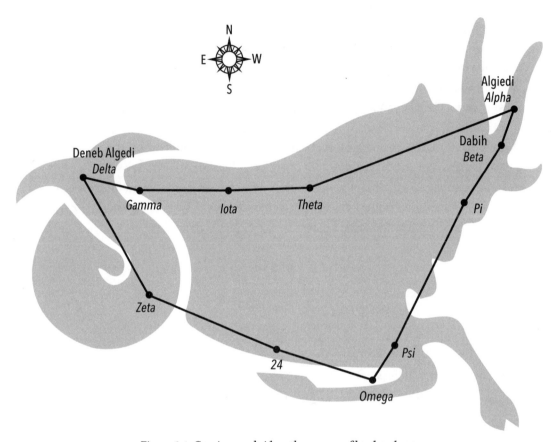

Figure 5.4. Capricornus bridges the powers of land and sea.

The Greeks associated the constellation with the forest deity Pan, who had the legs and horns of a goat. According to legend, Pan was honored with a place in the sky for helping to rescue the Olympian gods during their war with the Titans. During the battle against Typhon, Pan eluded the monster by jumping into a river and turning the lower part of his body into that of a fish so he could swim faster.

Another myth concerns the origin of the cornucopia or horn of plenty. This story tells how a goat, owned by the nymph Amalthea, suckled the infant Zeus. In gratitude, Zeus removed a horn from the goat and told Amalthea that it would provide her with anything she desired. Eventually, the horn was passed along to the river god Achelous, who used it to replace one of his own that had broken. Achelous was usually described as a mixture of land and marine animal.

The Tropic of Capricorn is an imaginary line south of the equator that marks the southernmost point at which the sun appears overhead at noon on the winter solstice. When this was noted two thousand years ago, Capricorn was in position as a backdrop to the sun and the line was named for this constellation. Because modern astrology uses the traditional dates for the zodiac, Capricorn is still the sign that begins winter. However, Sagittarius is the actual constellation in position on the other side of the sun on the solstice.

Ptolemy equated the stars of Capricornus with Mars and Venus, Mars and Mercury, or Saturn and Jupiter, depending on which part of the goatfish they depicted. In modern astrology, this constellation is equated only with Saturn. According to Nicholas Culpeper, Capricornus influences the bones, teeth, skin, and the right side of the body.

Notable Stars in Capricornus

Official Designation: Alpha Capricorni

Traditional Name: Algiedi

Pronunciation: al-JED-ee

Although Algiedi is the third brightest star in the constellation, its alpha designation is believed to come from its position as the westernmost star in Capricornus. In depictions, Algiedi is on the sea-goat's right horn. Also spelled Algedi, this star's traditional name, which is sometimes shortened to Giedi, comes from the Arabic word for "goat" or "kid." Algiedi is composed of two yellow stars known as Prima Giedi (Alpha-1) and Secunda Giedi (Alpha-2). These names mean "first and second Giedi."

Official Designation: Beta Capricorni

Traditional Name: Dabih

Pronunciation: DAH-bee

Located below Algiedi at the base of the goat's right horn, Dabih is actually two double stars. The first set of stars is called Dabih Major (Beta-1) and the second set is Dabih Minor (Beta-2). Beta-1a is orange and Beta-1b is blue-white. Dabih Minor is blue-white (Beta-2a) and white (Beta-2b). Beta Capricorni's traditional name, Dabih, comes from Arabic and means "the butcher." The reason for this is unknown.

Official Designation: Delta Capricorni

Traditional Name: Deneb Algedi

Pronunciation: DEN-ebb al-JEE-dee

Deneb Algedi is actually the brightest star in the constellation. Its traditional name comes from Arabic and means "the tail of the goat," which describes its location. It is frequently written with the alternate spelling of Algiedi. Deneb Algedi is a four-star system composed of two sets of white (Delta-1) and yellow-white (Delta-2) stars. Ptolemy equated Deneb Algedi with Jupiter and Saturn. Deneb Algedi is associated with wisdom. It was one of Agrippa's fifteen important fixed stars.

Magical Interpretations and Uses for Capricornus

This constellation has been linked with Pan and, quite naturally, the Horned God. As the Wheel of the Year turns, summer brings him into his prime, making this an excellent time to honor him as well as all men who follow a Pagan or Wiccan path. Whenever Capricornus is in the sky, it is a good time to hold a drumming circle to honor Pan, male fertility, and the wildness of nature. If possible, hold the circle at the beach to emphasize the sea-goat nature of Capricornus. Drumming and dancing is a great way to raise energy and get in touch with the wildness within our spirits. Chant the following as you drum or dance: *"Capricorn, hoof and horn; Wild spirit rise in me. Like a flame, be untamed; Capricornus, goat of the sea."*

In addition to marking the first harvest, Lughnasadh is a time to celebrate summer's abundance. This constellation's association with the horn of plenty makes the energy of Capricornus ideal to incorporate into this sabbat ritual. Since many of us decorate our altars or the area around them with fruits and vegetables, lay them out in the Capricornus star pattern. However, if space is limited you may want to opt for an image of the sea-goat on which you mark the star pattern. In addition, it's fairly easy to find a pendant with an image of Capricorn, which can be placed on your altar to help draw the energy of this constellation into your ritual.

Uniting the power of the sea and land, the sea-goat personifies the spirit of summer. Goats are playful and it is easy to imagine an animal with the combination of a goat and sea creature enjoying summer in the ocean waves. If you go to the beach, draw the constellation in the sand or lay out seashells in the star pattern of Capricornus. Connect with the playful energy of the sea-goat by chanting: *"Capricornus, goat of the sea; Bring delights of summer to me. With horns of abundance and tail of a fish; Bring pleasure and plenty, this is my wish."*

★ ★ ★

Cygnus: The Swan/Enchanted Shape-Shifter

Pronunciations: Cygnus (SIG-nus); Cygni (SIG-nee)

Visible Latitudes: 90° North to 40° South

Constellation Abbreviation: Cyg

Bordering Constellations: Aquila, Cepheus, Draco, Lyra, Pegasus

Description: The brightest stars in this constellation form an airplanelike shape as well as a
cross. What we may consider as the tail of the airplane is the long neck and head of the
swan pointing in a southwestern direction.

To Find: Locate the Summer Triangle. The star at the eastern anchor point is Deneb, which
marks the tail of the swan. At the center of the triangle is the star Albireo, which marks
the head of the swan.

This constellation is most frequently associated with the Greek myth of Zeus and Leda,
the Queen of Sparta, whom he seduced. Some versions of the story include Aphrodite
disguised as an eagle chasing Zeus, in the form of a swan, into the protective arms of
Leda. To commemorate his triumph in seducing the queen, Zeus placed an eagle and
swan in the heavens. Cygnus has also been regarded as Orpheus, the Greek hero and musician who was transformed into a swan after his death. Representing his lyre or harp, the
constellation Lyra is to the west of Cygnus. In some accounts, Zeus placed the lyre and
the eagle that rescued it among the stars.

A completely different myth concerns Phaeton, the mortal son of the sun god Helios,
and his friend Cycnus. Phaeton wrangled his father into allowing him to drive the chariot
of the sun. Helios warned him to stay on the circle of the zodiac, and although Phaeton
tried, he lost control of the chariot. The boys were thrown from the vehicle and fell to
earth. Cycnus survived but Phaeton fell into the River Eridanus. Cycnus called on Zeus to
turn him into a swan so he could dive into the water and recover his friend's body. Moved
by the boy's personal sacrifice to live the rest of his life as a swan, Zeus placed him in the
heavens where he could race among the stars forever.

The Arabs and Egyptians also regarded this constellation as a bird, but considered it a
hen. The Chinese associated it with a myth about a magpie bridge. In the story, the Goddess of Heaven forbade a fairy woman to marry a mortal man and created a river to keep
them apart. The bright ribbon of the Milky Way represented the river. According to the
legend, once a year all the magpies would come together to form a bridge (represented by
Cygnus) over the river, allowing the lovers to be together.

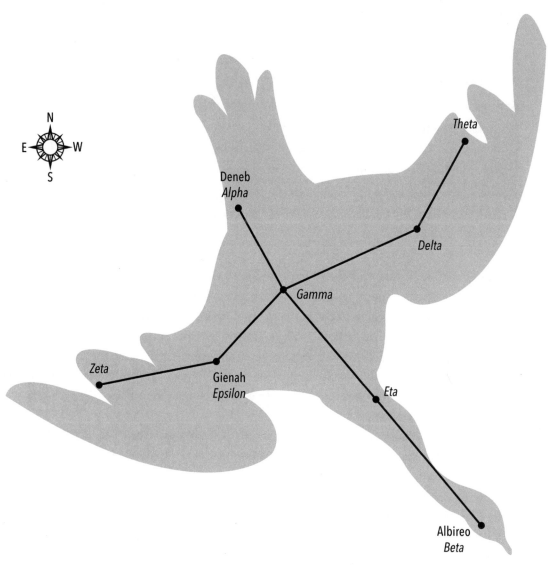

Figure 5.5. Cygnus provides a connection with enchantment and the power of prophecy.

Ptolemy equated the stars of Cygnus with Venus and Mercury. Through the ages this constellation has been considered highly mystical. In Hindu shamanic practice, it was common to wear a cloak of swan feathers while embarking on astral journeys.

Notable Stars in Cygnus

Official Designation: Alpha Cygni

Traditional Name: Deneb

Pronunciation: DEN-ebb

This luminous blue-white star marks the eastern anchor point for the Summer Triangle asterism. Its traditional name comes from Arabic and means "the tail," which describes its position on the swan. In the Chinese legend about the magpie bridge, Deneb represented the fairy chaperone who kept watch for the lovers.

Official Designation: Beta Cygni

Traditional Name: Albireo

Pronunciation: al-BEE-ree-oh

Albireo is a binary star with one yellow (Beta-1) and the other blue (Beta-2). Sometimes known as the Beak Star, Albireo marks the head of the swan. While it is not part of the Summer Triangle, it is easy to find because it is in the middle of this asterism.

Official Designation: Epsilon Cygni

Traditional Name: Gienah

Pronunciation: JEEN-ah

This orange star shares its traditional name with Gamma Corvi in the constellation Corvus the Crow. Gienah is Arabic and means "the wing."

Magical Interpretations and Uses for Cygnus

The swan is a symbol of light, death, transformation, and prophecy. It is associated with Aphrodite, Apollo, Venus, and the Muses. In Norse mythology, the swan symbolizes the soul. In Celtic tradition, it is associated with shape-shifting, enchantment, and otherworld beings. Linked with Bardic inspiration, it is often associated with the harp.

Although Beltane marks the time that fairies ride out from their *sidhe*, "fairy mounds," Midsummer's Night will forever be associated with them because of Shakespeare. Considered magical in their own right, swans are linked with the wee folk either as steeds or shape-shifted fairies. I like to call on the enchantment of Cygnus on Midsummer's Night for aid in acknowledging the fairies that inhabit my garden. I usually leave an offering of milk in a little swan-shaped vase. Acknowledging fairies can also be done by laying out the Cygnus star pattern with jade, peridot, or clear quartz as well as with clover, daisy, lavender, or rose flowers. Use this chant when you leave offerings: *"Cygnus, Cygni, flying so high; Call the fairies from your place in the sky. Let them know of these gifts I bring; Summer's here, time to dance and sing."*

Whenever I see swans paddling slowly on quiet water, it makes me think of legends about their magical songs having healing or sleep-inducing qualities. Of course, there's also the story of the ugly duckling, a tale that might be good for many of us to take to heart. Here's a meditation to try if you struggle with issues of self-esteem. On a starry summer night, clear floor space in the middle of a room where you will not be disturbed and gather eight pieces of clear or white quartz. Lay out the stones in the Cygnus pattern of stars, making it large enough for you to sit within the configuration under one of the wings. Fill a large, wide bowl with water and place it on the floor in front of you.

Lean forward until you can see your reflection in the water. Don't be shy about looking at yourself. Begin by saying: *"Oh, swan above, Cygnus, Cygni; Help find the beauty that lies within me."* Recall the story of the ugly duckling that felt so out of place, and then imagine your face morphing into that of a duckling. Think of the things that make you feel different, out of step, or awkward, but don't dwell on them; simply identify them. Now, imagine your reflection in the water turning into a graceful swan. Sit up straight with your head held high and close your eyes as you draw down the energy of the constellation. Your inner beauty and strengths are emerging as you come into your own power. Although you may always feel different or separate from others, remember that you are a swan and you hold your energy in the form of strength and beauty. Open your eyes, dip a finger in the water, and then draw the shape of Cygnus on your forehead as you say: *"Cygnus, Cygni, you have helped me to see; The powerful swan that exists within me."*

Draco: The Dragon/Wise Guardian
Pronunciations: Draco (DRAY-koe); Draconis (druh-KOE-niss)
Visible Latitudes: 90° North to 15° South
Constellation Abbreviation: Dra

Bordering Constellations: Boötes, Cepheus, Cygnus, Hercules, Lyra, Ursa Major, Ursa Minor

Description: Draco's pattern of stars winds between the two bears near the north pole. The dragon's head is defined by a triangle of stars.

To Find: Locate the line of stars that winds between the Big and Little Dippers. Follow them toward the bright star Vega in the Lyra constellation to locate Draco's head. Alternatively, locate Vega, and Draco's head is just to the northwest.

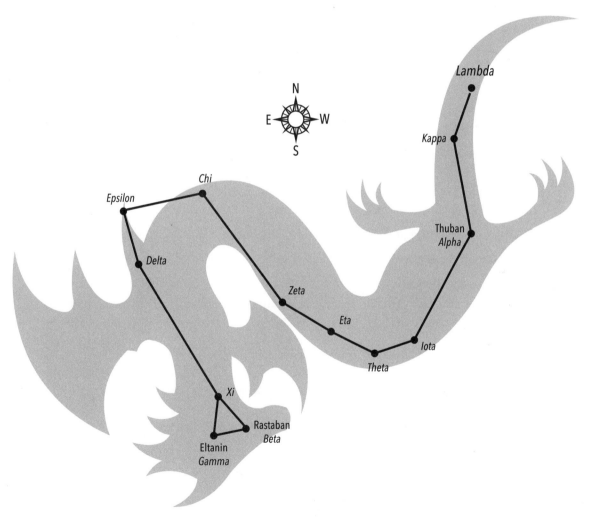

Figure 5.6. Controlling the power of the elements, Draco is the guardian of the earth.

Draco is a circumpolar constellation that does not set below the horizon for most observers in the Northern Hemisphere. Its name means "the dragon" in Latin and it represents Ladon, a dragon in Greek mythology. Ladon watched over the Hesperides (daughters of the god Atlas), who were the guardians of Hera's apple tree. Hercules was tasked with stealing some of the golden apples, which could only be accomplished after killing Ladon. To honor the slain dragon, Hera placed him among the stars.

In another legend, Draco was one of the Titans who fought many pitched battles against the gods of Olympus. The goddess Athena killed Draco and threw him into the sky. When the Romans told this story, it was Minerva who tossed the dragon into the sky.

The Egyptians regarded Draco as a crocodile, and the Babylonians associated it with Tiamat, the goddess of the primeval ocean, in the form of a sea serpent. The Anglo-Saxons referred to it as the Fire Drake. At one time, Ursa Minor was considered a wing on Draco's back.

According to Ptolemy, the stars of Draco were equated with Saturn and Mars. Additionally, the constellation is associated with skill, artistic abilities, and ingenuity.

Notable Stars in Draco

Official Designation: Alpha Draconis

Traditional Name: Thuban

Pronunciation: THEW-bahn

This white star is located near the end of the dragon's tail. Generally considered inconspicuous for an alpha, Thuban was given this designation because it was the pole star approximately 4,500 years ago. The star's traditional name comes from Arabic and means "the serpent."

Official Designation: Beta Draconis

Traditional Name: Rastaban

Pronunciation: RAS-tuh-bahn

The traditional name of this star also comes from Arabic and means "the serpent's head," which describes its location. Rastaban is a yellow star.

Official Designation: Gamma Draconis

Traditional Name: Eltanin

Pronunciation: EL-tah-nin

This orange star is the brightest in the constellation. During medieval times, Eltanin was known as Al Tinnin, from an Arabic word meaning "snake" or "serpent." Along with Rastaban it is located on the dragon's head.

Magical Interpretations and Uses for Draco

In the East, dragons are considered benevolent bringers of prosperity and good luck. In Western culture, the dragon has served as a prevalent emblem for some of the most-fierce warriors such as the Romans and the Vikings. To the Celts, the dragon was a symbol of sovereign power and today it graces the national flag of Wales. Along with serpents, the dragon came to represent Paganism, and images of Saint George slaying the dragon symbolized Christianity's triumph. However, to many twenty-first-century Pagans and Wiccans the dragon represents wisdom and power.

Along with their ability to fly and breathe fire, dragons were believed to live in the depths of the earth or in deep water. Symbolizing the four elements, they can help draw energy for ritual as well as power for magic circles. Following is an example for calling in the directions to build dragon energy:

"Soaring on the winds with the strength of your wing beats, draw the power of air to this circle, and be my guardian of the east.

With your mighty fiery roar, breathe the power of flame to this circle, and be my guardian of the south.

From the cool depths of lake and sea, bring the flow of water to this circle, and be my guardian of the west.

Within the darkness of your caverns, sovereign of ley lines, draw earth energy to this circle, and be my guardian of the north."

In many legends, dragons are guardians of great treasure hoards. Curling around the northern celestial region, Draco can be considered a guardian of the pivot point that turns our world. Along with the ability to control water, dragons guard and guide the energy of the land. Just as Hera enlisted the aid of a dragon to guard her tree of golden apples, we too can call on Draco to guard the things we grow and nurture in our gardens. To do this, place stones throughout your garden in the configuration of Draco's star pattern as you say, *"I call on thee, Draco, great dragon of the celestial sphere to connect your energy with the earthly dragons of the ley lines. As it is above, so be it below. Manifest the powers of soil and water in my humble garden. So mote it be."*

For protection of your home, on a starry summer night lay out Draco's star pattern with pieces of red zircon in a place where they won't be disturbed for a week. Light a stick of dragon's blood incense and draw the sign of a pentagram in the air over the gemstones. Walk through your home with the incense and as you enter each room, say: *"Draco, Draco, look down from on high; Send dragon power from your place in the sky. Surround and guard this home for me; Draco the dragon, may you my guardian be."*

Hercules: The Strongman/Dagda and Odin

Pronunciations: Hercules (HER-kew-leez); Herculis (HER-kew-liss)

Visible Latitudes: 90° North to 50° South

Constellation Abbreviation: Her

Bordering Constellations: Aquila, Boötes, Corona Borealis, Draco, Lyra, Ophiuchus, Serpens

Description: The star pattern has been described as forming the letter *H* for Hercules; however, it is easier to see the asterism called the Keystone, the square of stars that marks his lower torso. The figure is upside down in the sky.

To Find: Locate the bright star Vega in Lyra and the Keystone of Hercules is to the west. It is southwest of Draco's head.

Believed to have originated with the Sumerians and associated with Gilgamesh, the early Greeks called this constellation Engonasin, the Kneeling One. Later, Greeks identified this constellation with different heroes such as Prometheus, Ixion, and finally Hercules. The Phoenicians associated it with their god Melkarth.

Hercules was the son of Zeus and a mortal woman. Despite the name Hercules (Heracles in Greek) meaning "the glory of Hera," the wife of Zeus was enraged about her husband's dalliances with other women and made life as difficult as she could for Hercules. At one point, she made him go mad and during his insanity he killed his children. When asked for guidance, Apollo told Hercules that he could atone for the murders by serving Eurystheus, the king of Mycenae, for twelve years. As penance, Eurystheus gave him tasks that were considered impossible to accomplish. The constellation of the kneeling man was said to mark the event when Hercules knelt down and prayed to Zeus. Later star maps depicted him standing victorious with a foot on the head of the slain dragon represented by the constellation Draco.

Ptolemy equated this constellation with Mercury, and noted that it provides strength in character and determination in purpose.

Notable Stars in Hercules

Official Designation: Alpha Herculis

Traditional Name: Rasalgethi

Pronunciation: rah-sell-GAYTH-ee

Also spelled Ras Algethi, Rasalgethi is a multiple star system. Alpha-1a is orange with a blue-green companion (Alpha-1b). Alpha-2 is a binary star with one yellow (Alpha-2a) and the other yellow-white (Alpha-2b). The traditional name of this star comes from an Arabic phrase that means "the head of the kneeling one," which describes the star's location.

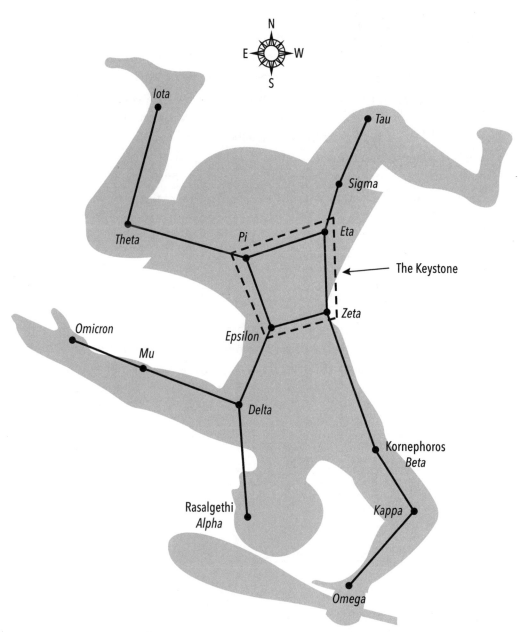

Figure 5.7. Hercules, representing the Dagda and Odin, brings strength and power.

Official Designation: Beta Herculis

Traditional Name: Kornephoros

Pronunciation: kore-neh-FOR-uss

Kornephoros is a yellow star and the brightest in the constellation. Its name comes from Greek and means "the club bearer." This star is located on Hercules's shoulder.

Magical Interpretations and Uses for Hercules

The Romans in Britain identified Hercules with the Celtic god the Dagda. The Dagda was a leader of the Tuatha Dé Danann and a principal god of Irish tradition. He was known for being a benevolent god and for possessing many skills. Like Hercules and several other gods, he was known for wielding a mighty club.

To many Pagans and Wiccans, the Dagda is a god of knowledge and wisdom as well as the source of potent creative energy. He can be called on at any time to boost the creative power of magic. With the constellation of Hercules in the night sky, the Dagda can be instrumental in helping us make contact with the fairy realm. On Midsummer's Eve, leave an offering outside for the fairies so they know you have friendly intentions. On Midsummer Night, walk in a circle to define your magical space and say: *"Hercules, Dagda, shining bright; On this starry Midsummer Night. I create a circle as I walk this ground; May the fairy mounds open and magic abound."* Sit in the middle of your circle and reach out with your energy to sense what is going on around you. Don't expect fireworks, as fairy contact may be very subtle, like the soft breath of a breeze or a slight shift in energy. Whether or not you detect contact, thank the fairies and leave an offering before dissolving your circle.

Well known as a strongman, Hercules can be called on for strength, especially during hard times. Throughout his story this hero of the ancient world was faced with many challenges. When we find ourselves in situations that seem overwhelming, we can turn to this constellation for energy to get us through the rough patches in our lives. Take a small piece of paper, a pencil, and a piece of string or ribbon, and sit outside on a starry night. Draw down the energy of Hercules, and then write a few keywords that describe the challenge or situation confronting you. Turn the paper over and draw the Hercules star pattern. When you are finished, roll up the paper so the star pattern is on the outside and tie it closed with a piece of string or ribbon so it looks like a little scroll. Hold it between your hands and review the challenge or difficulty for which you are seeking help. Follow this with a visualization of how you would like this to be resolved. Don't let fantasy take over; think in practical terms about how events could actually unfold or a situation be resolved. Leave the scroll outside for the night, or if there isn't a safe place to leave it, place it on an indoor windowsill. The next night, safely burn the scroll at your altar as you once again visualize the outcome of your challenge.

The epithets *Victor* "the winner" and *Invictus* "the undefeated" were used by cults devoted to Hercules. Celebrations were held in Rome on August 12 and 13 to honor him. These are ideal nights to draw on this constellation for strength in dealing with problems or for boosting your spells.

Because Hercules is upside down in the sky, I am reminded of Odin hanging from the world tree Yggdrasil as he sought wisdom. It was during this ordeal that he discerned the runes and brought their mystical meaning into the world. Call on Hercules before rune work or any type of divination to enhance your readings. Lay out the star pattern with runes or with the rune characters written on pieces of paper, and then draw down the energy of the constellation. This can also aid in opening your psyche for inspiration and to receive messages from other realms.

Lupus: The Wolf/Spirit Guide

Pronunciations: Lupus (LOO-pus); Lupi (LOO-pee)

Visible Latitudes: 35° North to 90° South

Constellation Abbreviation: Lup

Bordering Constellations: Ara, Centaurus, Libra, Scorpius

Description: The upper part of the star pattern looks like a crooked letter *H*.

To Find: Locate the Summer Triangle and from the star at its center (Albireo in Cygnus), draw an imaginary line toward the southwest to the red star Antares in Scorpius. Continue that line south and slightly west to the next constellation, Lupus.

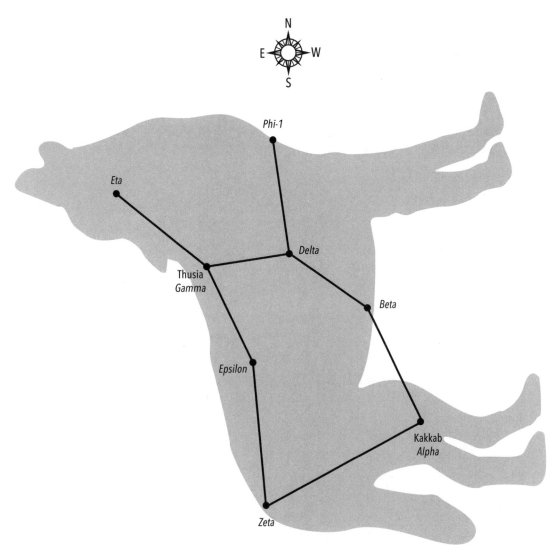

Figure 5.8. Lupus is a guiding spirit and guardian.

Lupus is a constellation of the Southern Hemisphere that can be seen in some parts of the north. Its name means "wolf" in Latin. To the Babylonians, the constellation was known as the Wild Dog, but to the Arabs it was the Lioness. These stars were originally considered part of the Centaurus constellation until Greek astronomer Hipparchus separated them and named the constellation *Therion*, "wild animal." Because of its proximity to

Centaurus and Ara, the Romans called it *Bestia*, "the beast," and interpreted Lupus as a sacrificial offering by the centaur Chiron to the gods.[22] The Lupus constellation became identified as a wolf and portrayed as such in Johann Bayer's 1603 star atlas.

According to Ptolemy, the stars in this constellation are equated with Saturn with a little bit of Mars energy.

Notable Stars in Lupus

Official Designation: Alpha Lupi

Traditional Name: Kakkab

Pronunciation: KACK-kab

This blue-white star marks one of the wolf's hind legs. The name Kakkab is not often used. It is believed to have come from a Babylonian phrase meaning "the star left of the horned bull," a reference to the constellation Centaurus, which they regarded as a bison-man.

Official Designation: Gamma Lupi

Traditional Name: Thusia

Pronunciation: thew-SEE-ah

The name of this blue-white star comes from Greek and means "sacrificial animal." Like the alpha star, the traditional name of this one is seldom used. In modern depictions of Lupus, this star is located on the wolf's back or neck.

Magical Interpretations and Uses for Lupus

The wolf is sacred to and associated with a number of gods and goddesses. Because of many misconceptions, this animal is associated with negative traits by some people. However, to others the wolf is a symbol of community, loyalty, protection, and spirit. It represents freedom and the power of the wilderness as well as discipline and the power of the group. It is a spirit animal and a powerful ally for psychic and shamanic work. If a wolf presents itself during astral travel, it will be your guide and guardian.

The Lupus constellation can help you call on the power of the wolf. The wolf is guided by instinct and can help you learn to trust your intuition. On a starry night, lay out the star pattern of Lupus on your altar with star glitter/confetti or tea light candles. Also, light at least one altar candle. Prepare yourself for energy work and then stand in front of your altar. Hold your arms out to the sides at shoulder height, palms facing upward as though you are going to hug the world, and howl like a wolf. It doesn't have to be loud, just allow yourself to feel the wildness and freedom of howling. Slowly bring your hands up overhead, and then bring your palms together. Lower your hands, keeping them together until they are in prayer position in front of your heart as you say: *"Lupus, Lupus, wolf of the sky; Be my guide as darkness draws nigh. Protect and lead me, while wisdom you share; My honor and loyalty to you, I swear."*

Sit in front of your altar, gaze at the candle flame for a minute or two, and then close your eyes. Bring the image of a wolf's face into your mind and imagine that you are looking into its eyes. Its deep gaze pulls you inward. Follow wherever the wolf leads because this powerful spirit can show you your place and function in the world. Through the wolf you will see your own power and learn to sense and trust your insights. Most of all, the wolf will help you come to know who you are and that you do not need to prove it to other people. Spend as much time as you need with the wolf and then let it go. You can always meet again. Leave the Lupus star pattern on your altar for a day or two to remind you of your encounter with this spirit animal.

Wolfsbane is a plant that is well known, but mainly from stories and films about werewolves. During medieval times it was associated with sorcery. Wolfsbane is the common name for *Aconitum lycoctonum* and it is sometimes applied to *Aconitum napellus*, which is more commonly known as monkshood. Blooming in mid to late summer, wolfsbane has soft yellow flowers on tall, upright spikes. The plant should be handled with care as it is poisonous and it should never be ingested. Wolfsbane grows in open woodlands and moist hillsides. If you find it growing wild, pick a few sprigs to take home to aid in drawing down the energy of Lupus. Even better, if you find it in a garden center, plant it in your yard where it can help draw the power of Lupus and represent the wild spirit of the wolf.

Lyra: The Harp/Otherworldly Guide
Pronunciations: Lyra (LIE-rah); Lyrae (LIE-rye)
Visible Latitudes: 90° North to 40° South

Constellation Abbreviation: Lyr

Bordering Constellations: Cygnus, Draco, Hercules

Description: A small parallelogram of four faint stars. Vega is northwest just outside of this shape depicting part of the instrument's handle.

To Find: Vega is the brightest star in the summer sky and is the western anchor point for the Summer Triangle asterism.

This constellation represents the lyre of Orpheus, the musician and poet of Greek mythology. A lyre is a type of small harp. According to some myths, Hermes invented it from a tortoise shell, which Apollo then gave to Orpheus. Orpheus was so talented that he enchanted the birds and tamed wild animals with his music. When his wife Eurydice was killed, he descended to the underworld to retrieve her. There, he charmed Pluto with his music and was allowed to leave with Eurydice, but only under the condition that they could not look back. As such stories go, temptation was too great and one glance back sealed their fates.

Another legend tells how the Thracian Maenads, female followers of Dionysus in northeastern Greece, murdered Orpheus because he had failed to honor the god of wine. According to one version of this story, his lyre was thrown into a river and Zeus, who sent an eagle to retrieve it, placed the instrument and the bird among the stars. Another version says that the Muses carried the lyre to heaven.

Lyra is sometimes associated with the Greek poet and musician Arion, whose melodies attracted a dolphin that ultimately saved his life. Delphinus represents the dolphin and this story is told in the entry for that constellation in chapter 6.

On old star maps, Lyra was often depicted as an eagle carrying the harp; however, it was sometimes depicted as just the bird itself. To the Arabs, Lyra was the Swooping or Falling Eagle; the Aquila constellation was their Flying Eagle. In India, Lyra was associated with the vulture instead of the eagle and was called the Falling Vulture. In the British Isles, the Anglo-Saxons and Celts considered Lyra as a harp and knew it as the Hero's Harp and King Arthur's Harp.

Ptolemy equated the stars of this constellation with Venus and Mercury. Through the ages Lyra has been associated with magic and spellwork.

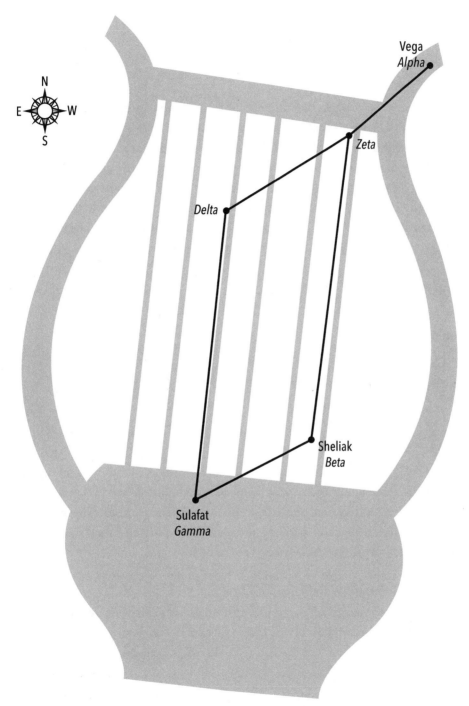

Figure 5.9. Lyra brings us sweet music from the heavens.

Notable Stars in Lyra

Official Designation: Alpha Lyrae

Traditional Name: Vega

Pronunciation: VAY-gah

Often spelled Wega, Vega is the fifth brightest star in the entire sky and the second brightest in the Northern Hemisphere. In depictions of Lyra, this blue-white star is located on the handle of the harp and was the northern pole star in approximately 12,000 BCE. To ancient Egyptians, this was the Vulture Star associated with the goddess Ma'at, who was often depicted holding a vulture feather. The name Vega is derived from the Arabic word "falling" or "swooping" in the name of the constellation. Vega also has the distinction of being the first star photographed in 1850. It is associated with artistic talents, social awareness, and magic. Vega was one of Agrippa's fifteen important fixed stars.

Official Designation: Beta Lyrae

Traditional Name: Sheliak

Pronunciation: SHELL-ee-ack

Sheliak is a blue-white spectroscopic binary star. Its traditional name is derived from Persian and means "the harp." It is usually depicted on the strings of the lyre.

Official Designation: Gamma Lyrae

Traditional Name: Sulafat

Pronunciation: SUEL-ah-faht

Sulafat is also blue-white and is actually the second-brightest star in the constellation. Its traditional name is derived from Arabic and means "the tortoise." This harkens back to the Greek story of Hermes inventing the harp from a tortoise shell. Like the beta star, Sulafat is usually depicted on the strings of the lyre.

Magical Interpretations and Uses for Lyra

Lyres and harps have served as ceremonial instruments since ancient times. Among the Norse, the harp represented a mystic ladder that connected heaven and earth. Harps were also said to become guides to the otherworld for their deceased owners, which is why they were usually placed on funeral pyres. Standard equipment for Celtic bards, harps could bring laughter, tears, or sleep. The Dagda's harp was said to hold his melodies until bidden to release them. In addition, the music of the Dagda's harp was said to initiate the change of seasons. To the Celts, harp music was associated with the otherworld and enchantment. This instrument also represents the immortality of the soul.

Summer is a special time because we can be outdoors in lightweight clothing that allows us to feel closer to the natural world. We can also spend more time outside at night and feel the enchantment of the dark. Use an iPod or other device with headphones or earbuds and sit outside on a starry night while you listen to harp music. Draw down the energy of Lyra, and then sit comfortably while you listen to one song several times until you can hold the melody in your mind. Turn off the device and remove the headphones or earbuds, but keep the harp music in your mind as you tune in to the energy of the natural world around you. The melody may begin to shift and change, but don't try to go back to the original tune. Instead, follow the altered song in your mind. Allow Lyra to guide you and you may encounter sweet music from the fairy realm.

Music touches us in a unique way, often more deeply than words. It can reach into our souls and help us move below the mundane façade of everyday life to find what is meaningful. Use the glitter/confetti or bluish gemstones (the stars of Lyra are blue-white) to lay out the pattern of stars on your altar. Gemstones such as blue lace agate or celestite/celestine are particularly good to use as they have an ethereal quality. Follow the previous exercise, but instead of removing the headphones or earbuds and opening to the world around you, keep listening to the music as you focus on yourself. Let the harp guide you inward to touch the essence of your soul. Let it also guide you to the appropriate time to end the session, and then sit in silence. Before moving on to other things, give yourself a big hug.

Music is also a form of celebration, and summer is a time to kick back and enjoy life. Lay out the pattern of Lyra's stars indoors or outside to serve as a reminder to slow down and listen to the music of life.

Ophiuchus and Serpens: The Serpent Bearer and the Serpent/Harmonious Energy

Pronunciations: Ophiuchus (ohf-ee-YOU-kuss); Ophiuchi (ohf-ee-YOU-kee)

Visible Latitudes: 60° North to 76° South

Constellation Abbreviation: Oph

Bordering Constellations: Aquila, Hercules, Libra, Sagittarius, Scorpius, Serpens

Pronunciations: Serpens (SIR-pens); Serpentis (sir-PEN-tiss)

Visible Latitudes: 74° North to 64° South

Constellation Abbreviation: Ser

Bordering Constellations: Aquila, Boötes, Corona Borealis, Hercules, Libra, Ophiuchus, Sagittarius, Virgo

Description: The somewhat dim stars of Ophiuchus form the shape of a large crystal pointing north and slightly east. Northwest of Ophiuchus is a triangle of stars that represents the head of Serpens. A line of stars to the east of Ophiuchus marks the serpent's tail.

To Find: Locate the bright star Vega, the western anchor point for the Summer Triangle. Draw an imaginary line southwest to Rasalhague, the star that marks the point of Ophiuchus's crystal shape. It is just south of Hercules.

Ophiuchus and Serpens are presented together in this book because they are an intertwined set of three constellations. Ophiuchus is in between the two parts of Serpens, which is the only constellation formed by two separate ones. Their individual names are Serpens Caput, meaning "the head of the snake," and Serpens Cauda, "the serpent's tail." Ophiuchus was formerly known by the name Serpentaurius. Its present name is derived from two Greek words that mean "serpent" and "to handle." In some ancient depictions, Ophiuchus is shown preventing Serpens from grabbing Corona Borealis, the Northern Crown.

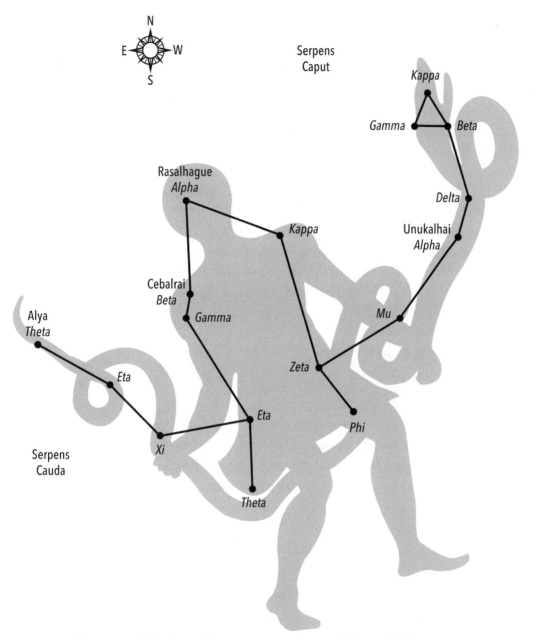

Figure 5.10. Ophiuchus and Serpens represent the eternal dance of life-force energy.

Originally, the Ophiuchus constellation represented Phorbas, the man hired by the people of Rhodes to rid their island of snakes. In later Greek myth, Ophiuchus represented the god of medicine, Asclepius, the son of Apollo. Legends vary as to whether Asclepius brought people back from the dead or that he was such a great healer few of his patients died. Either way, it caused a great deal of concern for Hades, god of the underworld, who was running short on customers. To solve his problem, Hades asked his brother Zeus to kill Asclepius. Zeus obliged and struck the god of medicine dead, but then put him in the sky as a lasting tribute to the healer's talents.

Before Greek influence, the Arabs considered this area of the sky as a pasture and Ophiuchus as a shepherd. As mentioned in chapter 1, Ophiuchus is the thirteenth constellation that falls within the ecliptic. Astronomically, Ophiuchus is noted for the number of new stars that seem to be born in its area of the sky.

According to Ptolemy, the Ophiuchus constellation is mostly equated with Saturn and a little with Venus. He considered Serpens to be equated with Saturn and Mars. In medieval medicine, Serpens was believed to aid in curing venomous snakebites.

Notable Stars in Ophiuchus

Official Designation: Alpha Ophiuchi

Traditional Name: Rasalhague

Pronunciation: rah-sell-HOG

Also spelled Ras Alhague, this double star marks the head of Ophiuchus. Its traditional name comes from Arabic and means "the head of the serpent charmer," a name that dates to post-Greek influence. Alpha-1 is a white star; Alpha-2 is orange.

Official Designation: Beta Ophiuchi

Traditional Name: Cebalrai

Pronunciation: SEB-all-rye

Cebalrai is an orange star that marks one of Ophiuchus's shoulders. Its name comes from Arabic and means "the shepherd's dog." This name dates to pre-Greek influence on Arab astronomy.

Notable Stars in Serpens

Official Designation: Alpha Serpentis

Traditional Name: Unukalhai

Pronunciation: uh-NOO-kool-eye

Unukalhai is an orange star located almost halfway between the head of the snake and Ophiuchus's body. Its name is from Arabic and means "the serpent's neck." In medieval Latin, this star was called Cor Serpentis, "the Serpent's Heart."

Official Designation: Theta Serpentis

Traditional Name: Alya

Pronunciation: al-you

Alya is a white binary star located near the end of Serpens's tail. Its Arabic name translates as "the sheep's tail." Like Cebalrai in Ophiuchus, the name dates to a time when the Arabs considered this part of the sky as a pasture and Ophiuchus as a shepherd.

Magical Interpretations and Uses for Ophiuchus and Serpens

These constellations can be instrumental in activating Kundalini energy. Running up our spines, this energy is portrayed as coiling snakes and appears much like the caduceus, the staff of Asclepius and the symbol of modern medicine. This relates to Ophiuchus portraying Asclepius, the god of healing, and Serpens, a snake. Before being cast as evil incarnate, snakes were regarded as agents of healing.

Kundalini energy is described as a sleeping serpent coiled around the first chakra at the base of the spine. When this energy is activated, it rises through two energy channels that weave back and forth across a central path along the spine. The goal of working with Kundalini is to move the full force of this energy upward, opening and activating all the chakras. Working with these channels along with the energy of these constellations can help us generate powerful personal energy for ritual and magic as well as healing.

Because the natural world helps us feel alive and balanced, doing this exercise outdoors is especially beneficial. Find a place where you can sit comfortably. Begin with your left hand over your solar plexus chakra and your right hand over your heart. Close your eyes and focus on the energy of these chakras until you feel them activate. Bring your attention down to your feet and the earth star chakra, and then draw earth energy up into your body. Since Kundalini energy is especially powerful, it is important to keep grounded.

Move your attention to the base of your spine. Imagine the energy rising on both sides of your spine. As it reaches each chakra, the energy flow crosses over to the other side of the body, weaving back and forth like the snakes of the caduceus. When your energy reaches the top of your head, visualize it rising higher through the three celestial chakras, meeting the descending energy of Ophiuchus and Serpens. Visualize this energy moving out across the area where you are sitting, spreading farther and farther, bringing healing and vitality to everyone and everything it encounters. Now visualize a little of that energy descending down through your chakras, bringing you balance, strength, and peace. Take time to feel your power, and then let your visualization gradually fade as you bring your awareness to your contact with Mother Earth. When you feel grounded, open your eyes and enjoy the beauty of the world around you and the sky above.

Sagittarius: The Archer/Wildness of Nature

Pronunciations: Sagittarius (sa-jih-TAIR-ee-us); Sagittarii (sa-jih-TAR-ee)

Visible Latitudes: 55° North to 90° South

Constellation Abbreviation: Sgr

Bordering Constellations: Aquila, Capricornus, Corona Australis, Ophiuchus, Scorpius, Serpens

Description: Although this constellation is often described as a teapot, the square in the upper part of Sagittarius's body is slightly easier to discern. This square is part of an asterism called the Milk Dipper. Also, a triangle of stars to the west represents the archer's bow.

To Find: Beginning at the Summer Triangle asterism, draw an imaginary line from Altair in Aquila at the southern anchor point of the triangle toward the tail of Scorpius. Sagittarius is two-thirds of the distance along that line. It does not rise far above the horizon and is not always visible in the Northern Hemisphere.

Sagittarius is a constellation of the Southern Hemisphere. Rooted in Sumerian myth, this constellation was adopted by the Babylonians, who depicted it on monuments. According to cuneiform inscriptions, it was called the Strong One. The Persians, Turks, and Syrians discerned either a bow or a bow and arrow in the star pattern. This constellation was also recognized in Egypt, and in India its name meant "the arrow."

Greek astronomer and mathematician Eratosthenes associated the constellation with Crotus, the son of Pan. Raised with the Muses, daughters of Zeus, Crotus was an accomplished musician and hunter as well as the inventor of archery. When he died, the Muses asked their father to honor him with a place among the stars. In classic Greek mythology, this constellation came to represent an archer who is usually depicted as a centaur holding a bow and arrow. A centaur is a creature with the upper body of a man on top of the body and four legs of a horse. This constellation is often wrongly identified as the centaur Chiron, who is represented by the constellation Centaurus. Sagittarius is usually portrayed aiming his arrow toward the heart of Scorpius, which is represented by the star Antares.

Ptolemy equated most of the stars in Sagittarius with Jupiter, along with Mars, Mercury, and Saturn, depending on their location in the constellation. He associated the arrow point with Mars and the moon. In modern astrology, Sagittarius is equated with Jupiter. According to Culpeper, Sagittarius influences the hips, thighs, sacrum, blood vessels, and pituitary gland.

Notable Stars in Sagittarius

Official Designation: Alpha Sagittarii

Traditional Name: Rukbat

Pronunciation: ROOK-baht

Rukbat is a blue star and although it is the alpha, it is not the brightest in the constellation. Its name comes from Arabic and means "the knee," which describes its location on the figure of Sagittarius. A variation in translation of the spelling of this name can be found in the delta star of Cassiopeia, Ruchbah. Rukbat is associated with steadiness.

Official Designation: Beta Sagittarii

Traditional Name: Arkab

Pronunciation: ARE-koob

Arkab is composed of multiple stars. Arkab Prior (Beta-1) is a double blue star. It is named Prior because it seems to lead Arkab Posterior (Beta-2) across the sky. Arkab Posterior is a yellow-white star. The name Arkab comes from Arabic and means "hamstring" or "tendon."

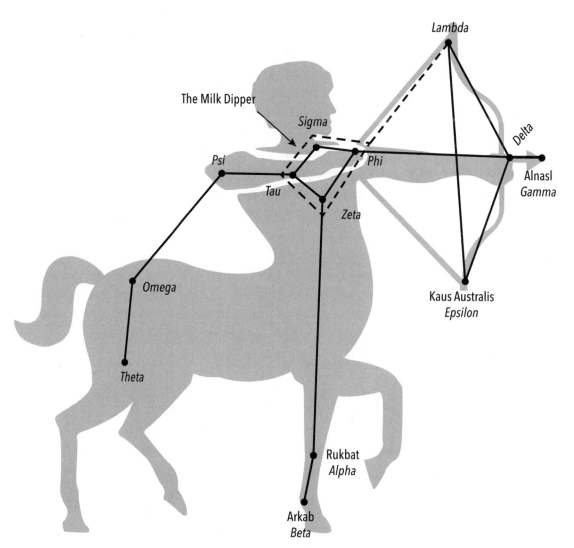

Figure 5.11. Sagittarius helps you take aim at your goals.

Official Designation: Epsilon Sagittarii

Traditional Name: Kaus Australis

Pronunciation: KOWS ow-STRAH-liss

Kaus Australis is a binary star, one blue (Epsilon-1) and one orange (Epsilon-2), and it is the brightest in the constellation. The name Kaus Australis is a mix of languages with the Arabic word for "bow" and the Latin word for "southern," respectively. This star is located at the base of the archer's bow and is associated with mental stimulation.

Official Designation: Gamma Sagittarii

Traditional Name: Alnasl

Pronunciation: al-NAH-zul

Alnasl is an orange star and its name means "arrow point" or "arrowhead." When we see this star we are looking toward the center of the Milky Way galaxy.

Magical Interpretations and Uses for Sagittarius

Like satyrs, centaurs symbolize male virility, stallion energy, and sexuality in general. They epitomize sensuality and aid in developing comfortable awareness of the physical body. With this in mind, sultry summer days are an ideal time for the following visualization. Find a place outside where you can sit without being disturbed. Imagine that you are standing in a meadow bordered by a forest. Looking down to the ground in front of you, you see two of your four hoofed feet. You are a centaur. As you become aware of the warm sun on your broad back, you swish your long, flowing tail, creating a slight breeze. Gazing across the meadow, you can see shimmering waves of heat rising skyward. Beyond and off to your right is a forest. The cool of the trees is inviting and you begin to move toward them; first at a slow pace, and then an easy trot. The movement feels invigorating, so you pick up speed to a full gallop. Air rushes past your face, and as you breathe harder you become aware of the fragrant plants in the meadow. And now, let your imagination take over to lead you into the forest or to turn around and enjoy the sun and the meadow.

Centaurs personify the general wildness of nature, but let's not forget the green world. Tidy gardens and manicured lawns show the influence of human hands, but beyond suburbia we find the wild glory of Gaia. Late July and early August are especially good times to observe the profusion of plants that echo the vitality of life itself. Take some berries along on a walk through a meadow or forest. If you know a place where berries grow wild, pick some along the way. When you find a place that seems appropriate for an offering, set out some of the berries in the Sagittarius star pattern as you say: *"In the name of Sagittarius, I leave this offering. May the beauty of his stars shine forever, and may my spirit know his wild freedom."*

Before guns and the advent of modern warfare, the most feared legions were the bowmen who could strike quietly from a distance. To be a bowman took a great deal of power and skill. With this in mind, an archer is the perfect symbol when you are planning changes in your life. Sagittarius reminds us to aim high and put our talents out for all to see. On a starry night lay out the pattern of the whole constellation or just the bow on your altar or any place where you will see it often. If possible, use blue gemstones such as azurite or lapis lazuli and one orange gemstone such as amber or sunstone for the star Alnasl at the point of the arrow.

Place something that symbolizes your goal opposite the arrow stone, and then pick up the orange arrow stone. Hold it between your hands and say: *"Let loose your arrow in the dark; Know my goal and find your mark."* Return the stone to the star pattern, and then hold the palms of both hands about an inch or two above the gemstones. Visualize the energy of Sagittarius coming through your celestial chakras, into your body, through your hands, and into the stones. Now, see that energy accumulate in the stone, representing the point of the arrow. Watch as the energy moves from the arrow toward the symbol of your goal and surrounds it. As this occurs, see yourself reaching that goal. Hold the image and feelings for a moment or two and then let them fade. Leave the stones and symbol of your goal in place for a day or two, or for as long as it feels appropriate.

Scorpius: The Scorpion/Death Wielder and Protector

Pronunciations: Scorpius (SKOR-pee-us); Scorpii (SKOR-pee-eye)

Visible Latitudes: 40° North to 90° South

Constellation Abbreviation: Sco

Bordering Constellations: Ara, Corona Australis, Libra, Lupus, Ophiuchus, Sagittarius

Description: This constellation actually resembles the creature it represents. A line of stars form the curving body of the scorpion, which ends in the curl of its poisonous tail.

To Find: Locate the Summer Triangle asterism and from the star at its center (Albireo in Cygnus), draw an imaginary line toward the southwest to the red star Antares, which marks the heart of Scorpius. The line of stars that trail to the southeast delineate the scorpion's tail. For those in the north, this constellation is low in the sky and for many of us it is not visible.

Located in the Southern Hemisphere, Scorpius is one of the oldest constellations. Five thousand years ago, the Sumerians considered it a scorpion. From approximately 5000 to 1000 BCE, it marked the autumn equinox and as such was considered a gateway to the otherworld as the dark time of year began. Bolstered by the scorpion's poisonous sting, the constellation remained associated with death and darkness.

In Greek mythology, this constellation was identified with the scorpion that killed Orion, the hunter. Opposite each other in the sky, the myth is played out with the Orion constellation setting as Scorpius rises. In one version of the story, Gaia sent the scorpion after Orion because of his boast that he could kill any wild beast.

During the period of ancient Greece, this constellation was considerably larger with two distinctive parts. The one we know today as Scorpius contains the body, tail, and stinger. What we know as Libra was the second part of Scorpius called Chelae Scorpionis, "the claws of the Scorpion." The Romans separated these constellations because they thought Libra looked more like weighing scales than scorpion claws.

Ptolemy equated this constellation mainly with Mars and Saturn, but also Jupiter or Mercury, depending on their location in the figure. In modern astrology, this constellation is equated with Mars and Pluto. According to Culpeper, this constellation influences the bladder, genitals, pubic bone, and nose.

Notable Stars in Scorpius

Official Designation: Alpha Scorpii

Traditional Name: Antares

Pronunciation: an-TAIR-eez

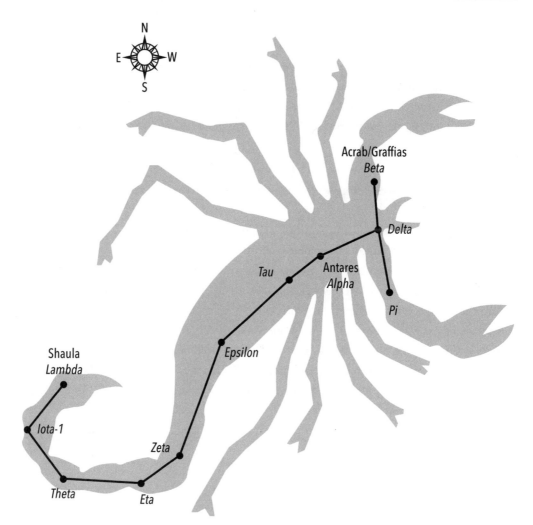

Figure 5.12. Scorpius is a fearsome protector.

This bright red star marks the heart of the scorpion. The name Antares comes from ancient Greek and has been translated as "anti-Ares" as well as "the rival of Mars." The latter is believed to refer to the star having a color similar to the planet Mars. The Babylonians called this star the Breast of the Scorpion, and its name in Latin, *Cor Scorpii*, means "scorpion's heart." In ancient Egypt, Antares represented the scorpion goddess Selket and was sacred to Isis. In Persia, Antares was one of the four royal stars and called the Guardian of the West. Additionally, it was one of Agrippa's fifteen important fixed stars.

Official Designation: Beta Scorpii

Traditional Names: Acrab; Graffias

Pronunciations: uh-KRAB; GRAFF-ee-us

Although only the sixth brightest in the constellation, its prominent position gave this blue-white binary star its beta designation. The name Acrab, which is also spelled Akrab, comes from Arabic and means "the scorpion." Its other name, Graffias, is derived from a Greek word meaning "crab." In depictions of the scorpion it is located at the base of one of the claws.

Official Designation: Lambda Scorpii

Traditional Name: Shaula

Pronunciation: SHOW-lah

Despite its designation deep into the letters of the Greek alphabet, Shaula is actually the second-brightest star in the constellation. Its traditional name comes from Arabic and means "the scorpion's stinger," which describes its location on the tail. Shaula is a blue-white triple star system.

Magical Interpretations and Uses for Scorpius

The scorpion is a small but potent creature that has had evil or negative connotations in many cultures, but not all. To the Egyptians and Babylonians, it was a symbol of protection. The goddess Selket, protector of the dead, was depicted with a scorpion on her head, and Isis escaped from the murderous Seth under the protection of seven scorpions. In addition, the Babylonians believed that scorpion-people guarded the seven gates of the underworld. If you are feeling in need of protection, carry or wear a Scorpio pendant to attract good luck and repel negativity. To prepare it as an amulet, dab it with peppermint or rosemary oil, and draw down the energy of this constellation as you say: "*Scorpius, Scorpius, bring luck to me. Protect and remove negativity.*" Alternatively, instead of using oil on the pendant you could pass it through the smoke of frankincense.

Scorpius is a backdrop to the sun in late November during the dark of the year. Its lesson during the long days of summer is to embrace your dark side. Darkness is the place of incubation for creativity, spirituality, transformation, and psychic skills. When engaging in activities to explore deeper aspects of the self and/or develop psychic abilities, let Scorpius give your efforts a boost. Before you begin your activity, lay out the constellation's pattern of stars using bloodstone, labradorite, and/or moonstone as you say: *"Scorpius, Scorpius, whom many fear; In dark of night I call you near. Help me draw forth my talent and skill; So mote it be, this is my will."*

Chapter Six

THE AUTUMN QUARTER OF SEPTEMBER, OCTOBER, NOVEMBER

September marks a shift in the daily routine and mindset. Summer is over, yet many afternoons are still bright and warm. While nights start to turn chilly, winter's threat seems a far horizon as autumn colors begin to blaze with breathtaking beauty. Mabon marks the autumn equinox, and like its spring counterpart, it is a day of balance between light and dark. As we gather in the last of the summer harvest, we also draw spiritual abundance into our lives. We count our blessings and give thanks to the Lord and Lady.

A mild breeze whispers on sunny October afternoons. This is a time to pause and enjoy the magic and beauty of the earth, but change is in the air. Night comes earlier as the harbingers of winter steal leaves from the trees and scatter them crumpled and brown on the ground. Moonlight shimmers on the first wet frost as the earth prepares to rest and incubate the next cycle. November brings us into the dark of the year. After confronting our own mortality at Samhain, this month invites us to turn inward for reflection, study, and to further our skills.

The autumn night sky is called the Cassiopeia Quadrant because of this constellation's distinctive shape that makes it an easy marker to find and then locate other stars. As a circumpolar constellation, Queen Cassiopeia can be seen throughout the year. However, she is most visible now and, quite appropriately, presides over this season when we give thanks to the Goddess for the bounty of the earth. Known to the ancient Greeks as the Sea, the autumn night sky is a watery place. It is populated by two fish, a dolphin, a whale, and a water bearer. Elsewhere in this quadrant we will find a ram, a winged horse with a colt, and a queen with her king and princess daughter. Let's see what a Pagan perspective can do with these.

As a reminder, some of the bordering constellations noted throughout this chapter may fall within other seasons. The directions given to locate constellations and stars assumes that the reader is facing south. Also, have a look at chapter 8 as some of the southern constellations listed only in that chapter may be visible to you.

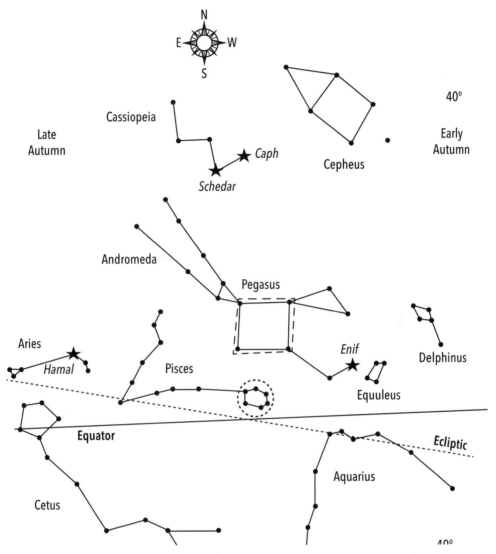

Figure 6.1. The autumn night sky. The dotted shapes show the Great Square of Pegasus and the Circlet of Pisces asterisms. Four of the brightest stars are also noted.

THE AUTUMN CONSTELLATIONS

Andromeda: The Princess/Horse Goddess

Pronunciations: Andromeda (an-DRAH-meh-duh); Andromedae (an-DROM-uh-die)

Visible Latitudes: 90° North to 40° South

Constellation Abbreviation: And

Bordering Constellations: Cassiopeia, Pegasus, Perseus, Pisces

Description: Two gently curving lines of stars form a slanted, elongated V shape.

To Find: Andromeda appears below her mother, Cassiopeia. However, the easiest way to find her is to locate the Great Square of Pegasus asterism. The northeastern corner is Andromeda's alpha star, Alpheratz. This is located at the bottom of Andromeda's V shape and represents her head.

This constellation is believed to predate the classical civilizations of Greece and Rome, however, its true origin is unknown. While it is named after a mythical princess, the constellation is sometimes called the Chained Woman; Persea, "wife of Perseus"; or Cepheis, "daughter of Cepheus."

In Greek mythology, Andromeda was the daughter of King Cepheus and Queen Cassiopeia. The extremely vain queen angered the Nereids (sea nymphs) by boasting that she and her daughter were more beautiful. In a rather overblown act of revenge for the slander, Poseidon, the nymphs' father, sent a sea monster to flood and destroy their land. The sea monster is represented by the constellation Cetus. Advised to sacrifice his daughter to the monster in order to save his country, the king had her chained to a rock at the edge of the sea. Andromeda was resigned to her fate, but in true fairy-tale fashion, Perseus swooped in on the winged horse Pegasus and saved her. The goddess Athena honored this heroic deed by placing them among the stars. In fact, most of the characters in this drama populate the northern region of the sky.

According to Ptolemy, the stars in this constellation are equated with Venus. In medieval medicine, this constellation was said to engender love in marriage and to reconcile problems with adultery.

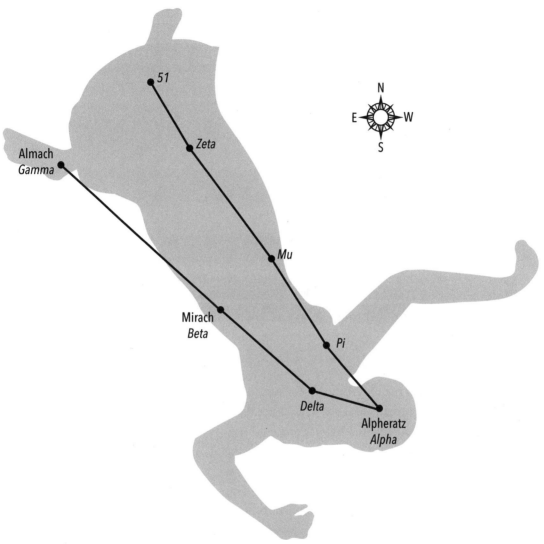

Figure 6.2. Andromeda represents the restoration of feminine power.

Notable Stars in Andromeda

Official Designation: Alpha Andromedae

Traditional Names: Alpheratz; Sirrah

Pronunciations: AL-fer-rats; SIR-ah

Alpheratz is a blue binary star that represents Andromeda's head. Both of its traditional names are derived from an Arabic phrase that means "the horse's navel," referring to Pegasus. Connecting two constellations, this star marks the northeastern corner of the Great Square of Pegasus asterism. At one time it was also designated as the delta star in Pegasus, but that has been dropped. Along with the stars Caph in Cassiopeia and Algenib in Pegasus, it was known as one of the Three Guides. This star is associated with independence and freedom.

Official Designation: Beta Andromedae

Traditional Name: Mirach

Pronunciation: MIRR-ahk

This red star is located midway on the constellation's V shape and marks Andromeda's left hip. Its traditional name is derived from an Arabic word that means "the girdle" or "the waist-cloth." Mirach is associated with feminine power and intuition.

Official Designation: Gamma Andromedae

Traditional Name: Almach

Pronunciation: ALL-mock

Almach is located at the top of the V shape near Andromeda's left foot. Totally unrelated to the story of Andromeda, this star's name comes from Arabic and means "a small predatory animal," which is believed to refer to some type of wild cat local to the Middle East.[23] The orange Gamma-1 is the brightest component of Almach. Gamma-2 is blue, and Gamma-3 is a white spectroscopic binary star. This star is associated with artistic abilities.

Magical Interpretations and Uses for Andromeda

With this constellation's alpha star, Alpheratz, marking a corner of the Great Square of Pegasus, Andromeda is forever linked with the horse. Like goddesses such as Epona, Rhiannon, and Macha, she shares power with this animal. While Andromeda started out as a chained and helpless woman, Pegasus, representing the forces of nature, ignited her personal power. We can call on the energy of Andromeda for aid in breaking our chains and removing anything that holds us back from gaining or maintaining our personal power. In addition, this constellation can be called upon for spells of release.

Gather a few leaves that have fallen to the ground and dried. Oak, apple, hazel, or acacia leaves are good choices because they are associated with letting go. Light a black candle on your altar, place a piece of paper in front of it, and then put the leaves on the paper. Using a black felt-tip pen, write a few keywords on one of the leaves that describes a fear, something you want to release from your life, or something that is preventing you from developing your personal power. Draw the star pattern of the Andromeda constellation across the keywords. Use a second and third leaf if necessary for other things you want to release. When you are finished, hold the leaves between your hands and say: *"Andromeda, Andromeda, please draw near; I hold in my hands the things I fear. Help me release them and throw away strife; May hope and beauty enter my life."* Crumble the leaves onto the paper, and then fold it over them, making a little package. Blow out the candle and leave the package on your altar for the night. The next day, take it to a place in nature, ideally a pond or river, and scatter the bits of leaves on the water as you repeat the incantation. When you return home, burn the piece of paper or tear it into small pieces and bury it.

It seems appropriate that the Andromeda constellation is almost directly overhead on Samhain. Her legend tells of her facing death, and at this sabbat we take time to acknowledge the fact that we, too, will step over that threshold someday. The following can be included in your Samhain ritual if you work solo or it can be done separately when you are alone any night either side of the sabbat. Use eight pomegranate seeds to lay out Andromeda's star pattern on your altar. As you do this, say: *"Dark Mother, as I stand at the gateway between the worlds this night, may the light of Andromeda shine on me. May she help me see that if I fear death I cannot fully live, and if I fear life I will not find solace in death. When my time comes, may Andromeda on her winged horse carry me peacefully over that threshold and show me the way into the otherworld beyond. So mote it be."* One by one, eat the pomegranate seeds from the star pattern as you contemplate the wonder, beauty, and delicacy of life and the cycles that form the ongoing spiral from one generation to another.

Aquarius: The Water Bearer/Sea God
Pronunciations: Aquarius (uh-KWARE-ee-us); Aquarii (uh-KWARE-ee-eye)
Visible Latitudes: 65° North to 90° South
Constellation Abbreviation: Aqr

Bordering Constellations: Aquila, Capricornus, Cetus, Delphinus, Equuleus, Pegasus, Pisces, Piscis Austrinus

Description: A gentle zigzag line of stars forms a crooked, upside-down letter *U*.

To Find: Start at the Great Square of Pegasus and the star Alpheratz (the head of Andromeda) at the northeast corner of the square. Draw an imaginary line to the star Markab at the opposite, southwest corner of the square. Continuing that line will bring you to Sadalmelik, the alpha star in Aquarius.

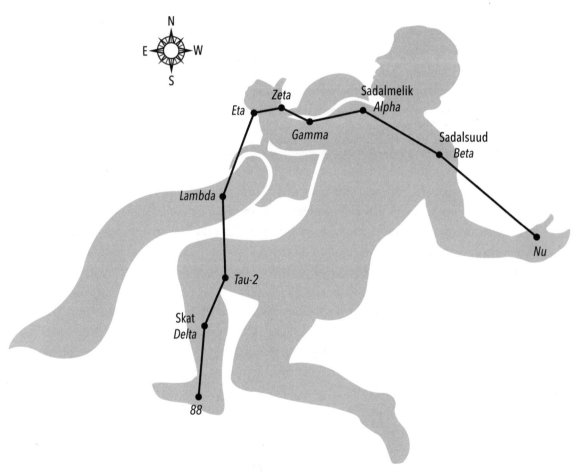

Figure 6.3. Aquarius carries the power and magic of water.

The name Aquarius is Latin and means "the water bearer" or "the cupbearer." Located in the area of the sky that the Greeks called the Sea, Aquarius fits in with a number of other constellations that relate to water. In fact, up until the second century BCE Aquarius was considered part of the constellation Eridanus the River.

Aquarius is sometimes depicted as a man pouring water from a large jar into the mouth of a fish, which is represented by the constellation Piscis Austrinus the Southern Fish. In addition, Aquarius is often associated with Ganymede in Greek mythology. He was a beautiful boy whom Zeus, in the form of an eagle, abducted to serve as cupbearer to the gods. The springtime constellation Crater is sometimes regarded as Ganymede's cup.

Other legends associate Aquarius with Deucalion, the son of Prometheus, who survived the great deluge. The Sumerians and Babylonians regarded Aquarius as their gods Enki and Ea, respectively. Associated with water and creation, these gods epitomized the power of the Tigris and Euphrates Rivers. Aquarius became equated with abundant water because the sun rose in this constellation during the rainy season in the Mediterranean region. The Egyptians associated it with Hapi, god of the Nile River and bringer of life-giving water. Aquarius was sometimes depicted as pouring water into the Nile and measuring its rise.

Ptolemy equated the stars of this constellation with Saturn and Mercury. In modern astrology it is equated only with Saturn. Culpeper credited this constellation with influencing the lower legs, ankles, circulation, and pineal gland.

Notable Stars in Aquarius

Official Designation: Alpha Aquarii

Traditional Name: Sadalmelik

Pronunciation: sah-dool-MEL-ik

Sadalmelik is a yellow star that marks one of the water bearer's shoulders. Its name comes from an Arabic phrase that means "the lucky star of the king."

Official Designation: Beta Aquarii

Traditional Name: Sadalsuud

Pronunciation: sah-dull-su-OOD

This yellow star is the brightest in the constellation and marks Aquarius's other shoulder. The name Sadalsuud comes from Arabic and means "the luck of lucks." It also has the Latin name of *Lucida Fortunae Fortunarum*, "the brightest luck of lucks" or "the luckiest of the lucky." Both the alpha and beta stars are associated with attracting positive events and energies.

Official Designation: Delta Aquarii

Traditional Name: Skat

Pronunciation: SCOT

Skat is also a yellow star with a name coming from Arabic. It is most often interpreted as "leg" or "shin"; however, it may have been derived from a word that means "a wish." The last meaning seems appropriate because the other stars in the constellation are associated with luck.

Magical Interpretations and Uses for Aquarius

Aquarius's location in an area of the sky called the Sea, surrounded by other constellations depicting ocean creatures, brings a sea-god quality to this constellation. Just as the Sumerians and Babylonians associated him with Enki and Ea, and the Egyptians with Hapi, he could equally represent Poseidon, Neptune, Manannán, Aegir, or Njord. Aquarius carries the power of the oceans and water in general.

In approximately 4000 BCE, Aquarius was the background constellation for the sun during the winter solstice. Even though it no longer holds that position at the solstice today, it still represents beginnings and is famous for marking the dawn of a great new age. The opposite of beginnings is, of course, endings; however, Aquarius's appearance in the night sky during the autumn and at Samhain is less about the ending of a cycle and more about preparing for the next. Summer and the blazing fall colors have passed, but winter is not yet upon us. Aquarius reminds us that this is a time to prepare for winter and get ready to nurture the new cycle that is to come. Just as the death aspect of the Great Mother Goddess is always linked with regeneration and renewal, the dark of winter is a time of incubation for whatever we want to develop. Whether you want to expand creative talents, psychic abilities, or magical skills, call on Aquarius to help you prepare mentally and energetically for these endeavors.

Prepare your altar by laying out the Aquarius star pattern with gold glitter/confetti, citrine gemstones, or small pieces of yellow paper. You will also need a small jar of water and an empty bowl that is large enough to hold the water from the jar. Light a yellow candle. Sit in front of your altar and as you draw down the energy of this constellation, say: *"Aquarius, Aquarius, shining this night; Pour down energy and golden light. Help me prepare, develop, and grow; Nurture the seeds in the spring I will sow."* Repeat this incantation as you pour the water from the jar into the bowl and visualize Aquarius pouring golden light and energy on you. When the jar is empty, hold the image of golden light surrounding you for another moment or two, and then let it fade. Before blowing out the candle, say: *"Aquarius, god of the celestial sea, may your water of golden light prepare and nourish my endeavors. Blessed be."*

Since the stars in this constellation are associated with luck, the energy of Aquarius can give your good luck spells a boost. Prepare a yellow candle by carving the Aquarius star pattern or the names of the stars into it. Below this, carve a word or two relating to the reason you are seeking luck, and then anoint the candle with jasmine or spikenard oil. When the oil is dry and before lighting the candle, gently touch the end of it to each of your shoulders and a shin, the locations of the three notable stars on depictions of Aquarius. As you do this, say: *"Lucky stars, yellow and bright; Shine on me this beautiful night."* Place the candle in its holder and light it. Visualize what you want to achieve and then say: *"Lucky stars, yellow and bright; Shine on me this beautiful night. I wish I may, I wish I might; Have luck for what I have in sight."* Blow out the candle and visualize the smoke rising to Aquarius, taking your wishes with it.

Aries: The Ram/Power of the Horned God

Pronunciations: Aries (AIR-eez); Arietis (air-ee-ah-tiss)

Visible Latitudes: 90° North and 60° South

Constellation Abbreviation: Ari

Bordering Constellations: Cetus, Perseus, Pisces, Taurus

Description: The three brightest stars form an obtuse angle on the ram's head. Another triangle of stars marks its rear flank.

To Find: Start at the Great Square of Pegasus and the star Alpheratz (the head of Andromeda) at the northeast corner of the square. Follow Andromeda's stars northeast up to
the top of the *V* and then draw an imaginary line to the southeast.

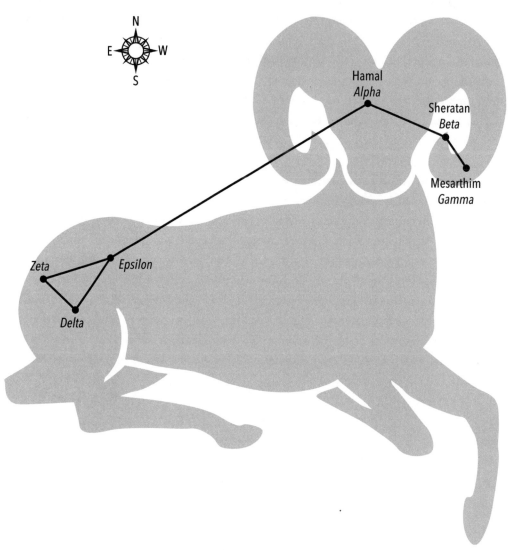

Figure 6.4. Aries is the god of abundance and strength.

Aries is the first sign of the zodiac in modern astrology because when the system was devised this constellation marked the vernal equinox. However, Pisces is the actual constellation in position on the other side of the sun on the equinox.

Associated with the mythological golden ram, this constellation's name is Latin and means "the ram." The legend is about a king in a region of Greece and his second wife, the quintessential evil stepmother, who wanted to do away with her husband's offspring. The children were a boy and girl, Phrixus and Helle. Just in the nick of time before the stepmother could strike, a winged ram with a golden fleece arrived to save them. As they flew away, Helle fell off and drowned in the strait of water in Turkey that was called the Hellespont in her honor. It is now known as the Dardanelles.

When Phrixus arrived at a place of safety, he sacrificed the ram and presented it to King Aeëtes. It may seem like a cruel end for the creature that had saved his life, but the act illustrated his gratitude to the powers that sent the ram to his rescue. In another legend, this animal's pelt became the quest object for Jason and the Argonauts. The Aries constellation has also been regarded as Zeus disguised as a ram to escape the Titan giants during their epic battle with the Olympian gods.

According to Ptolemy, most of the stars in this constellation are equated with Mars and Saturn, a few with Mercury, and others with Venus. To modern astrologers, Aries is equated with Mars. According to Culpeper, this constellation influences the head, eyes, face, upper jaw, and front of the body.

Notable Stars in Aries

Official Designation: Alpha Arietis

Traditional Name: Hamal

Pronunciation: HAHM-al

Hamal is an orange star and originally it was the only one that marked the celestial ram. Its name comes from Arabic and means "sheep" or "lamb," which is how the constellation was interpreted. Although it is usually depicted on the ram's head, Ptolemy called this star the One Above the Head. Hamal is associated with action and independence.

Official Designation: Beta Arietis

Traditional Name: Sheratan

Pronunciation: SHARE-ah-tahn

Sheratan is a binary star with a white primary component and a yellow companion. This star's traditional name comes from an Arabic phrase that has two interpretations: "the two signs" and "the two attendants." The name is believed to refer to a time when the beta and gamma stars seemed to herald the vernal equinox.

Official Designation: Gamma Arietis

Traditional Name: Mesarthim

Pronunciation: mess-AHR-tim

This is a triple star system that includes a binary star (both components are white) and one orange star. The name Mesarthim has the same origin as Sheratan, the beta star, and was one of the two stars that heralded the vernal equinox. The Persians called Sheratan and Mesarthim the Protecting Pair. Both stars are usually depicted on one of the ram's horns.

Magical Interpretations and Uses for Aries

The ram is associated with the Horned God and serves as a symbol of fertility and abundance. The Egyptian solar god Amun is associated with this constellation, as is Khnum, the Egyptian god of creation, who was depicted with a ram's head. As a symbol of strength, the ram is also associated with Ares, Belenus, Hermes, Pan, Marduk, and Mars. Call on the power of Aries for October or November dark moon observances or any ritual in which you want to call on the power of the god. Lay out the Aries star pattern on your altar using agate, bloodstone, or carnelian. Call down the energy of Aries to cast your circle, saying: *"Lord of abundance and strength, join me in creating this circle that echoes the turning shape of a ram's horn. May the power of your presence guide and protect all who stand within."*

Because Aries is associated with the terrestrial animal and the Horned God, this constellation offers a high degree of protective energy. To invite this presence to your property, paint the star pattern on a rock that you will place outdoors. Choose a spot where you feel protective energy is needed or anywhere for general protection. On a starry night, put the rock in place as you say: *"Aries, Aries, I call to thee; Bring protective power to me. Let it flow into this stone; With your grace, protect this home."* Apartment dwellers can make adjustments by placing the rock in a potted plant, on a windowsill, or on a balcony. Walk around your property or apartment three times as you repeat the incantation.

Another way to draw down the power of this constellation is to walk a spiral as described in chapter 3. This is especially appropriate when working with Aries as the spiral echoes the winding shape of a ram's horn. You will need enough floor space to lay out a spiral and a couple of strands of garland or a rope to create it. Start at the center and allow enough space to sit there. As you wind the rope or garland outward, make the space in between each turning wide enough to walk. Before you begin your walk into the spiral, pause at the entrance to ground your energy and focus on your purpose for calling on Aries. Once you begin, say this as a mantra: *"As I walk this spiral tonight; Aries lend your power and might."*

When you reach the center of the spiral, draw down the energy of Aries. Visualize your purpose and repeat the Aries mantra three times. Spend as much time as feels appropriate and then slowly walk out of the spiral as you say: *"Aries lend your power and might; For what I seek this starry night."*

Cepheus and Cassiopeia: The King and Queen/The Lord and Lady

Pronunciations: Cassiopeia (kass-ee-oh-PEE-uh); Cassiopeiae (kass-ee-oh-pee-eye)

Visible Latitudes: 90° North to 20° South

Constellation Abbreviation: Cas

Bordering Constellations: Andromeda, Cepheus, Perseus

Description: Five stars in this constellation form a distinctive letter *W*.

To Find: Locate the Great Square of Pegasus and draw an imaginary line north. The two bright stars mark the right side of Cassiopeia's W shape.

Pronunciations: Cepheus (SEE-fee-us); Cephei (see-fee-EYE)

Visible Latitudes: 90° North to 10° South

Constellation Abbreviation: Cep

Bordering Constellations: Cassiopeia, Cygnus, Draco, Ursa Minor

Description: Five of the stars in this constellation form a connected square and rectangle that resembles a child's drawing of a house.

To Find: Locate the W of stars in Cassiopeia. From the bright star at the bottom right (west) of the W shape, draw an imaginary line to the star at the top right of the W. Continue that line and you will come to the faint square of stars in Cepheus.

Cepheus and Cassiopeia are circumpolar constellations and can be seen throughout the year sitting on their thrones. I have presented them together because their legends as well as their roles for Pagans and Wiccans are intertwined.

In Greek mythology, Cepheus was the king of Ethiopia, and Cassiopeia was his wife and queen. This kingdom of Ethiopia is not the modern-day country in Africa. Its location is thought to have been a stretch of land along the Mediterranean and the Red Seas that encompassed parts of present-day Egypt, Jordan, Israel, and Lebanon.

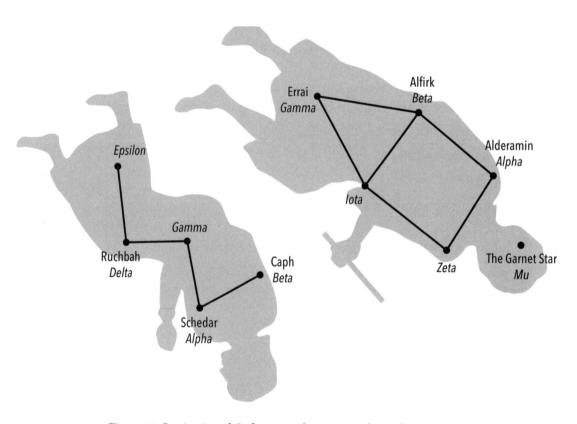

Figure 6.5. Cassiopeia and Cepheus provide power together and separately.

Both Cassiopeia and her daughter Andromeda were exceptionally pretty, but the queen was extremely vain and boasted that they were more beautiful than all of the Nereids. This caused a ruckus with the sea nymphs, who complained to their father, Poseidon. To avenge their honor, Poseidon sent a sea monster to flood and destroy Ethiopia. King Cepheus was advised that the only way to save his country was to sacrifice his daughter to the sea monster. Luckily for Andromeda, Perseus saved her and the monster was slain. While Andromeda and Perseus were placed in the sky by Athena to honor the daring rescue, Poseidon placed Cassiopeia near the celestial pole, where she was condemned to spend half of the year upside down. Zeus placed the king among the stars because Cepheus was descended from one of his lovers, the nymph Io.

The constellation of Cepheus is believed to predate Greek mythology, and according to author and astrologer Bernadette Brady, Cepheus was an "important image of male sovereignty"[24] that portrayed strength in balance with the Great Goddess. While the Greek story of Cassiopeia belittles female power, her constellation was noted on Assyrian tablets as the vitally important Lady of the Corn. The constellation was also known as the Creatress of Seed. Also, the Arabs regarded Cepheus as a shepherd.

Ptolemy equated Cassiopeia with Saturn and Venus, and Cepheus with Saturn and Jupiter. Cassiopeia is associated with the ability to command respect, and Cepheus with authority. In medieval medicine, the constellation of Cassiopeia was believed to help restore strength to weak bodies.

Notable Stars in Cassiopeia

Official Designation: Alpha Cassiopeiae

Traditional Name: Schedar

Pronunciation: SHED-er

Schedar is an orange star located at the bottom right of the W in the star pattern. This star's traditional name is Arabic and means "the breast," which describes its location on the star figure.

Official Designation: Beta Cassiopeiae

Traditional Name: Caph

Pronunciation: KAFF

This yellow-white star is the brightest in the constellation. Its traditional name is derived from an Arabic phrase that means "the stained hand," which refers to the use of henna for the ritual tattooing of hands and feet. Caph is located at the top right of the W in the star pattern. Along with the stars Alpheratz in Andromeda and Algenib in Pegasus, Caph was known as one of the Three Guides.

Official Designation: Delta Cassiopeiae

Traditional Name: Ruchbah

Pronunciation: ROOK-baht

Ruchbah is a blue-white binary star. Its name comes from Arabic and means "the knee." A variation in translation of the spelling of this star's traditional name can be found in the alpha star of Sagittarius, Ruckbat. Ruchbah is located at the bottom left of this constellation's W shape.

Notable Stars in Cepheus

Official Designation: Alpha Cephei

Traditional Name: Alderamin

Pronunciation: al-DER-ah-min

This white star marks the king's right shoulder. Its traditional name comes from an Arabic phrase that means "the right arm." If you look at Cepheus as an upright house shape, Alderamin would be at the lower, right corner.

Official Designation: Beta Cephei

Traditional Name: Alfirk

Pronunciation: all-firk

The brightest component of this triple star, Beta-1, is blue and its two companions are white. The name comes from Arabic and means "the flock," left over from a time when Cepheus was regarded as a shepherd. In an upright house shape, Alfirk would be the upper right corner.

Official Designation: Gamma Cephei

Traditional Name: Errai

Pronunciation: err-RYE

Marking the king's knee, Errai is an orange binary star. Its traditional name is derived from Arabic and means "the shepherd." In the Cepheus house shape, Errai is the peak of the roof.

Official Designation: Mu Cephei

Traditional Name: The Garnet Star

This luminous red star is one of the largest in the Milky Way. It is often called Herschel's Garnet Star for German-born astronomer Sir William Herschel (1738–1822) who was so delighted with its color that he brought it to worldwide attention in 1783. This star is outside of and below the house-shape pattern. On the star figure of Cepheus it is located on his head.

Magical Interpretations and Uses for Cassiopeia and Cepheus

With their position at the top of the celestial dome and their myth as king and queen, Cassiopeia and Cepheus represent the god and goddess, the Lord and Lady. They preside over the natural world in balance, especially at the autumn equinox. Not only do they bring balance with male and female energies, but they also turn the cycles of life and death, light and dark. The Welsh name for the Cassiopeia constellation is Llys Dôn, "the Court of Dôn." Dôn is the Welsh mother goddess and the equivalent of the Irish Danu. Beli or Beli Mawr is the father god of the Welsh and Dôn's husband. He is the god of the dead and she is the life-giving mother. At Samhain, incorporate Cassiopeia and Cepheus into your ritual or simply take some time to call on the energy of these constellations as you remember those you love who have passed beyond the veil. Call down the energy of these constellations by saying: *"As the world passes into the dark of the year with Cassiopeia and Cepheus in the heavens above, I call to Dôn and Beli, the Lord and Lady to preside over this sacred circle. Beli, god of the dead, receive those I love who have passed into your realm this year."* Take time to name them and then say: *"Lady Dôn, when the time is right, give them renewed life. So mote it be."*

During autumn backyard cleanup, use stones to lay out the star patterns of Cassiopeia and Cepheus in your garden so the Lord and Lady can preside over your property. In the spring and summer these stones will disappear under a canopy of plants, but in the autumn they will re-emerge and be visible again. This will also serve as a reminder of the turning Wheel of the Year as well as the eternal presence of the god and goddess.

Cassiopeia and Cepheus not only represent the Lord and Lady, but also the relationships of lovers. Call on the energy of these constellations for spells in finding your true love and for adding spark to an ongoing union. Additionally, with the stars of Cepheus portraying an image of a house, call on the energy of his constellation for matters dealing with home and family.

On her own, the power of Cassiopeia can be instrumental in healing or dealing with mother/daughter relationship issues. For this you will need a six-inch square of white cloth, a felt-tip marker, approximately eight inches of red ribbon or yarn, a piece of moonstone, and a piece of bloodstone. As you assemble these things, burn a little myrrh to clear the energy of the room. Use the marker to draw Cassiopeia's star pattern in the center of the cloth. As noted in the description, the major stars of this constellation form the letter *W*, which can also be the letter *M* if you look at it the other way around. For this purpose think of it as *M* for mother. Place the two gemstones over the star pattern, gather the cloth up around them, and tie it closed with the ribbon, making a little sachet. Hold it between your hands and say: *"Cassiopeia, mother above; Help me sort out this complicated love."* Think of how you want to resolve the issues with your mother or daughter as you draw down the energy of the stars through you and into the sachet. Tuck the sachet into your purse so you can carry it with you, especially when you are with your mother or daughter. Of course, this can extend to relationship issues with mothers- and daughters-in-law.

Cetus: The Whale/Keeper of Traditions
Pronunciations: Cetus (SEE-tus); Ceti (SEE-tee)

Visible Latitudes: 70° North to 90° South

Constellation Abbreviation: Cet

Bordering Constellations: Aquarius, Aries, Eridanus, Pisces, Taurus

Description: A circle of stars marks the whale's head. Other stars trail down toward the southwest and end with his split tail. Most of the stars in Cetus are faint.

To Find: Begin at the *W* of Cassiopeia. From the star at the top left (east) of the *W*, draw an imaginary line south and slightly east between Aries and Pisces. The whale stretches southwest under Aries and Pisces toward Aquarius.

Derived from the Greek word *ketos*, Cetus is Latin for "whale." Along with other water creatures, this constellation is in the region of the sky that the ancients called the Sea. Cetus was usually depicted as a hybrid creature described as a dog with a mermaidlike tail. In the Greek legend of Andromeda, Cetus was the monster to whom the princess was to be sacrificed. The hero, Perseus, who rescued Andromeda, used the head of Medusa to defeat the monster. Medusa was the snake-haired Gorgon whose glance, even in death, turned onlookers to stone.

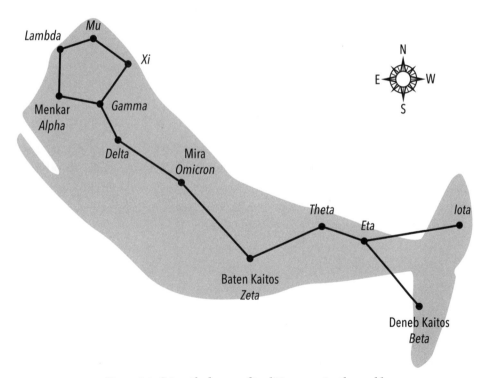

Figure 6.6. Cetus, the keeper of traditions, carries the world.

It was natural for the Greeks to consider this constellation as a sea monster because it had been regarded as a type of leviathan by earlier stargazers. To the Babylonians it represented the chaotic and frightening unknown of deep water. In seventeenth-century Europe, this constellation became associated with the Christian story of Jonah and the Whale, and Cetus remained an ominous, scary creature.

Ptolemy equated the stars of Cetus with Saturn. According to author and astrologer Bernadette Brady, this constellation is associated with unconscious forces.[25]

Notable Stars in Cetus

Official Designation: Alpha Ceti

Traditional Name: Menkar

Pronunciation: MEN-car

The name of this red star is Arabic and means "nose," although in many depictions it is located lower on the creature's head. If Menkar's scientific designation seems familiar, it is because Alpha Ceti has been a popular name with science-fiction writers.

Official Designation: Beta Ceti

Traditional Names: Deneb Kaitos; Diphda

Pronunciations: DEN-ebb KAY-tohs; DIFF-duh

Deneb Kaitos is an orange star and the brightest in the constellation. Its name comes from an Arabic phrase that means "the southern branch of the monster's tail." Its other name, Diphda, means "the second frog." At one time it was associated with the star Fomalhaut in Piscis Austrinus and both were regarded as frogs. Fomalhaut was called the First Frog because it seemed to lead Beta Ceti across the sky.

Official Designation: Omicron Ceti

Traditional Name: Mira

Pronunciation: MEE-ruh

Mira is a binary star of which one component is red (Omicron-1) and the other white (Omicron-2). In 1596, the German astronomer David Fabricius (1564–1617) noted this star's changeable characteristics, making it the first variable star to be discovered. Its traditional name is Latin and means "the amazing one" or "wonderful." It was so named because, as variable stars do, it seems to disappear and reappear as its brightness varies.

Official Designation: Zeta Ceti

Traditional Name: Baten Kaitos

Pronunciation: BAH-tun KYE-tuss

Although in most depictions this star is located on the side of the whale, its traditional name means "the belly of the sea monster" in Arabic. Baten Kaitos is an orange star.

Magical Interpretations and Uses for Cetus

Luckily we have moved beyond the stories of Jonah and of Moby Dick to regard whales quite differently. Nowadays instead of devilish monsters, they are considered majestic giants and a symbol of the importance of living in harmony with the natural world. Revered by Native Americans of the Pacific Northwest, whales are often depicted on totem poles. According to the legends of various traditions, the earth rests on the back of a great creature, usually a tortoise or whale, that guides our planet safely on its course through the sky. The whale is a symbol of great power and Cetus represents a power that is larger than us. It holds us and lifts us up.

Call on the power of Cetus when you feel that you need to be lifted through a rough period in your life. Draw the Cetus star pattern on a piece of paper, and then place it under your bed or under your mattress. Before going to sleep, say: *"Cetus, in the celestial sea; Lift me up and carry me. From this difficulty set me free; Oh mighty whale, blessed be."* Do this each day until your challenge begins to pass, and then remove the Cetus drawing from under your bed. Light a candle on your altar, thank Cetus, and then safely burn the paper.

As people have learned in the last several decades, whales are highly intelligent creatures. They are keepers of traditions, which they communicate through complex musical patterns. Whales are animals of the underworld fearlessly diving deep into the darkness. Yet these amazingly large creatures are powerful enough to rise up out of the water to momentarily soar, embodying the phrase "as above, so below."

For Mabon, carve the Cetus star pattern on a candle to represent the balance of the equinox. Use the energy of this constellation to cast your ritual circle, saying: *"In sweet majestic harmony; Cetus swims the celestial sea. Enfold the world with your song; Help to make this circle strong. Cetus, with your power vast; This magic circle now is cast."* Also use this when casting circles for spellwork.

Fifty-five million years ago the whale's rugged and adaptable ancestors, like other marine mammals, left the changing environment on land for life in the sea. High above in the celestial sea, Cetus reminds us that we can always make changes in our lives. Try this to help send your energy toward the changes you seek. Cut open an apple, take out the seeds, rinse them in water, and let them dry. On a starry night, hold the seeds in your hands as you draw down the energy of Cetus and visualize the change you want to bring into your life. Release the energy into the seeds, and then use them to lay out the Cetus star pattern on your altar. Leave the seeds in place for three days, and then bury them in the ground. Apples are associated with transformation, and burying the seeds is a symbolic act that carries the intention of change and growth.

Delphinus: The Dolphin / Carrier of Souls

Pronunciations: Delphinus (dell-FIE-nuss); Delphini (dell-FIE-nee)

Visible Latitudes: 90° North to 70° South

Constellation Abbreviation: Del

Bordering Constellations: Aquarius, Aquila, Equuleus, Pegasus

Description: A small, kitelike shape is formed by a diamond of four stars with a fifth star below to the southwest suggesting the tail of a kite.

To Find: Start at the Great Square of Pegasus. An imaginary line drawn from the center of the square westward would pass north of Equuleus to Delphinus.

This constellation's name is Latin and means "dolphin." Like a number of other celestial marine animals, Delphinus is located in the area of the sky the ancients called the Sea. According to one Greek legend, this constellation represents the dolphin that acted as Poseidon's messenger or go-between while he was courting the sea nymph Amphitrite. The sea god honored the dolphin by placing an image of him among the stars.

Another legend concerns a dolphin that saved the life of poet and musician Arion. When Arion was sailing back to Greece after visiting Italy, the sailors plotted to kill him and take his money. Having noticed dolphins swimming alongside the ship whenever he played his lyre, Arion requested that the sailors allow him to sing one more song before he

died. As he played, the dolphins swam alongside the ship and the musician jumped over-
board onto one of the creatures, which took him all the way to Greece. To commemorate
the rescue, Apollo placed the dolphin among the stars and called it *Vector Arionis*, "victorious Arion." Apollo also made a constellation of Arion's harp, which is represented by
Lyra. In other legends, Apollo, who was an accomplished lyre musician, transformed himself into a dolphin and was known as Apollo Delphinios.

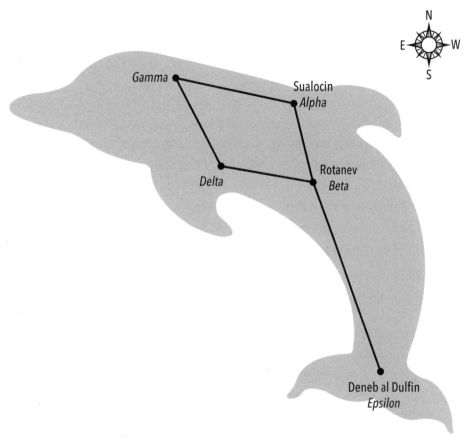

Figure 6.7. Delphinus carries souls to the otherworld.

While this constellation was regarded as a dolphin in India, the Arabs regarded it as
a camel. Ptolemy noted that the stars of Delphinus were equated with Saturn and Mars.
Delphinus is associated with cheerfulness and a fondness for travel.

Notable Stars in Delphinus

Official Designation: Alpha Delphini

Traditional Name: Sualocin

Pronunciation: SWAH-low-sin

The primary component of this multiple star system is blue-white. Sualocin and Rotanev, the beta star, were named by and for Italian astronomer Niccolò Cacciatore (1770–1841). Using the Latinized version of his name "Nicolaus Venator," he spelled his first name backward to name the alpha star and his last name backward for the beta star.[26]

Official Designation: Beta Delphini

Traditional Name: Rotanev

Pronunciation: ROE-tuh-nev

Rotanev is a yellow-white star. Along with Sualocin, Rotanev marks the back of the dolphin. Both stars are associated with talent.

Official Designation: Epsilon Delphini

Traditional Name: Deneb al Dulfin

Pronunciation: DEN-ebb al DULL-fin

Deneb al Dulfin is a white binary star. The name is Arabic and means "the tail of the dolphin," which describes its location on the star figure.

Magical Interpretations and Uses for Delphinus

Dolphins are friendly, playful, extremely intelligent, and social animals. They are the focus of a great deal of scientific study because they seem interested in and reach out to humans. Their wise and otherworldly qualities have fascinated people for thousands of years. Dolphins have served as a symbol of transformation, beauty, fun, and the sea. Since ancient times, mariners believed that dolphins could warn them of approaching storms. Call on the energy of Delphinus for support in weather and other forms of divination. Like dolphins, this constellation can aid in working with elemental magic, releasing negativity, and discovering truth.

To the Greeks, Romans, and Celts, dolphins personified the journey to the Isle of the Blessed. In Greek legend, it was Apollo Delphinios who carried souls to the land of the dead. Greek coins depicting dolphin riders often included a cockleshell beneath the dolphin. The famed Celtic artifact known as the Gundestrup cauldron also portrays a rider on a dolphin, which has been interpreted as a liberated soul.[27] Ferrying the departed to the otherworld combined with the association of transformation makes Delphinus appropriate for Samhain. Draw the star pattern of Delphinus on the inside of a cockleshell, scallop, or clamshell. Place it on your altar to represent the passing over of loved ones, or as part of your Samhain ritual hold it in your hands as you draw down the energy of Delphinus while saying: *"Hail dolphin Delphinus, ferryman of souls; Glide across the heavens, seas, and shoals. Carry those I love and keep them safe; Until I once more behold each face."*

Dolphins have been a symbol of pregnancy because of the similarity in the Greek words *delphis*, "dolphin," and *delphus*, "womb." If you are pregnant or planning a pregnancy, call on the power of Delphinus during the darkness of a new moon to aid with fertility as well as protection during pregnancy. Prepare a blue candle with jasmine oil or burn jasmine incense, and lay out the Delphinus star pattern using bloodstone, hematite, red jasper, or moonstone. Put your hands on your belly as you draw down the energy of this constellation while saying three times: *"May the baby in this womb; Come to term, grow, and bloom. Delphis, Delphinus swimming above; Protect this baby that I so much love."* Leave the stones in place until the full moon, and then move them to the room where your baby's crib is or will be located.

As in the legend of musician Arion, dolphins carry travelers safely to their destinations. With this association in mind, use the energy of the Delphinus constellation for travel spells. You will need a cockleshell, scallop, or clamshell and pieces of aquamarine, blue zircon, or moonstone to represent each person who will be traveling. On a starry night, draw the star pattern of Delphinus on the inside of the seashell and place the stone(s) in it. Hold the shell in front of you as you walk through your home, saying: *"Dolphin above, guide and protect me; As I travel o'er land, air, or sea. Keep me safe wherever I roam; Then bring me safely back to my home."* Place the shell with the gemstone(s) on your altar until you return from your travels.

Pegasus and Equuleus: The Winged Horse and the Colt/Powers of Nature

Pronunciations: Pegasus (PEG-ah-suss); Pegasi (PEG-ah-see)

Visible Latitudes: 90° North to 60° South

Constellation Abbreviation: Peg

Bordering Constellations: Andromeda, Aquarius, Cygnus, Delphinus, Equuleus, Pisces

Description: The Great Square of Pegasus represents the main body of the horse. It is a prominent asterism made up of three bright stars in Pegasus and one in Andromeda. This constellation represents the front half of the horse's body.

Pronunciations: Equuleus (eh-KWOO-lee-us); Equulei (eh-KWOO-lee-eye)

Visible Latitudes: 90° North to 80° South

Constellation Abbreviation: Equ

Bordering Constellations: Aquarius, Delphinus, Pegasus

Description: Equuleus is a small, faint constellation. Four of its most visible stars create a parallelogram that represents the colt's head, neck, and shoulder.

To Find: Locate the Great Square of Pegasus. Equuleus is west and slightly south of the square.

The Pegasus constellation is known for its Great Square asterism as well as for its number of bright stars. There is no definitive explanation as to why this star figure depicts only half of the horse. However, according to writer William Olcott, this was intended to convey the idea that Pegasus is soaring through the clouds.[28]

Greek mythology has several stories about the creation of the white, winged horse. One explains how this immortal offspring of Poseidon and Medusa sprang from her neck when Perseus beheaded her. Medusa was the snake-haired Gorgon whose glance turned onlookers to stone. Another legend says that when Medusa's head was severed, drops of blood fell into the sea. From her blood and sea foam, Pegasus emerged. The name Pegasus comes from a Greek phrase that means "the spring of the ocean."[29] The winged horse was said to have been tamed by either Neptune or Minerva.

Other stories about Pegasus tell of his flying to the home of the Muses on Mount Helicon, where after striking his hoof against the ground the Hippocrene, "horse's fountain," was formed. According to legend, anyone who drank from the fountain was gifted with the ability to write poetry. Pegasus was also said to have assisted Zeus by carrying his thunderbolts.

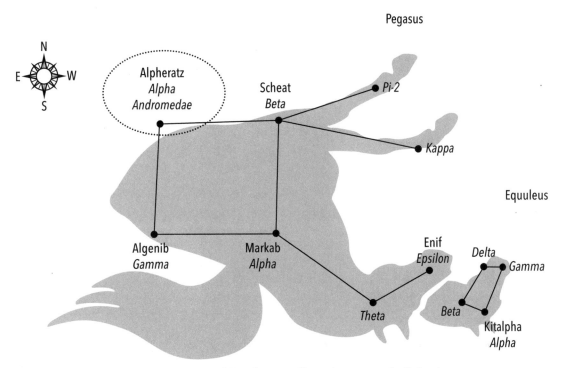

Figure 6.8. *Pegasus and Equuleus provide ancient powers for divination and guidance for travel to other realms.*

Vedic myths from India tell of a horse called Pagas, which was represented by the alpha and beta stars of Aries. Ptolemy indicated that the stars of Pegasus were equated with Mars and Mercury. Pegasus is associated with action, ability, and knowledge. In medieval medicine, this constellation was called upon for aid in preventing diseases in horses as well as to free a person from the influence of witchcraft.

The name Equuleus is Latin and means "the little horse" or "foal." It is one of the smallest of the eighty-eight constellations. Equuleus has been called the Foal of the Heavens, and because it precedes Pegasus across the sky, it was known as *Equus Primus*, "the first horse."

The Greeks associated this constellation with Hippe, the daughter of the centaur Chiron. Her name means "mare." She became pregnant after being seduced by Aeolus, a god of the wind. Too ashamed to tell her father, Hippe hid in the mountains until her child was born and prayed to the gods that Chiron would not find her. According to legend, Artemis answered her plea and put Hippe and her child in the heavens. Hippe was associated with prophecy. According to another legend, Equuleus was the foal Celeris, which was given to Castor as a gift from Mercury. He represented the offspring of Pegasus.

Notable Stars in Pegasus

Official Designation: Alpha Pegasi

Traditional Name: Markab

Pronunciation: MAR-kab

This blue-white star at the lower right (southwest) corner of the Great Square is usually depicted on the horse's wing. Its traditional name comes from Arabic and means "the horse's saddle." Markab is associated with stability.

Official Designation: Beta Pegasi

Traditional Name: Scheat

Pronunciation: SHEE-aht

This red star marks the upper right (northwest) corner of the Great Square. Its traditional name comes from Arabic and means "the shin," although in most depictions it is at the top of the horse's leg. Scheat is associated with creativity.

Official Designation: Gamma Pegasi

Traditional Name: Algenib

Pronunciation: al-JEN-ihb

Algenib is a blue-white star located at the lower left (southeast) corner of the Great Square. Also derived from Arabic, its name means either "the side" or "the wing." Along with the stars Caph in Cassiopeia and Alpheratz in Andromeda, Algenib was known as one of the Three Guides.

Official Designation: Epsilon Pegasi

Traditional Name: Enif

Pronunciation: EEN-if

Enif is an orange star and the brightest in this constellation. Its traditional name comes from an Arabic word that means "the nose," and as you might expect, it is located on the horse's muzzle.

Notable Star in Equuleus

Official Designation: Alpha Equulei

Traditional Name: Kitalpha

Pronunciation: kit-AL-fah

This yellow, spectroscopic binary star is the only one in the constellation that has a traditional name. Kitalpha comes from an Arabic phrase that means "part of the horse." It marks the neck or shoulder of the colt.

Magical Interpretations and Uses for Pegasus and Equuleus

Throughout the world, the horse was represented as coming from the darkness of the earth or sea, and springing forth with abundant life-force energy. Befitting the balance of the autumn equinox, the duality of the horse represents both life and death. The White Horse of Uffington on the Berkshire Downs in England is a stylized representation of a horse created by cutting away turf to reveal the glistening white chalk beneath. It has been a ritual site since ancient times and indicates the importance of this animal. Goddesses such as Epona, Rhiannon, Macha, and others were intimately associated with the horse and its power.

A winged horse is a symbol of the ability to fly to heaven or journey to the underworld. In particular, a white horse represents a bridge for transversing realms. In Celtic myth, the sea god Manannán's horses were said to be visible amongst the white, curling wave tops. His horses could carry people across the water as well as to the otherworld under the sea.

Call on Pegasus to carry you over the threshold at Samhain to bring you into contact with loved ones on the other side of the veil. With this constellation's proximity to the celestial sea, Pegasus is a natural vehicle and guide for astral travel. On a terrestrial level,

Pegasus helps us cross the threshold from one season into the other as winter approaches. Whatever threshold you are about to cross, on a starry night lay out the Pegasus star pattern on your altar with star glitter/confetti or pieces of white quartz. Light a black candle and then draw down the energy of Pegasus. State your purpose and then say: *"Through the air or under the sea; Winged horse above, carry me. At the threshold that I must go through; Pegasus, Pegasus, I call to you."*

In Irish folklore, horses were believed to have second sight and the ability to see ghosts. Additionally, the Celts and Germanic tribes practiced forms of divination involving horses. Because of this, I think of the Great Square of Pegasus as a window into other realms with the power to aid us in divination and other psychic practices. The association of horses with protection enhances the power that Pegasus can lend for these endeavors. Before you begin a divination session, lay out the constellation's star pattern using pieces of clear quartz, white coral, black agate, jet, or black tourmaline in any combination. Place your divination tools within or on top of the square. Sprinkle a pinch of dried herbs such as black cohosh or coltsfoot over your tools and the star pattern. Draw down the energy of Pegasus and release it into your divination tools and the gemstones.

Autumn is the time when children go back to school. This event is a major milestone, whether children are just beginning school or going off to college. Although it is a proud day for parents, it can also be a little difficult to experience the separation from our children that this initiates. Of course, no matter where or how far away they go from us, we always remain connected. As is most often the case, our children are fine and we are the ones who need reassurance.

On the night before your child starts school or moves out of the house, lay out the Pegasus and Equuleus star patterns on your altar using blue lace agate, citrine, jade, or rose quartz in any combination. Draw down the energy of both constellations as you say: *"Pegasus and Equuleus; My child I will soon miss. Keep him/her safe as he/she goes away; He/she will always in my heart stay. Although his/her time away will seem long; Our love and bonds are forever strong."* Leave the star pattern in place for a day and then put the stones away.

Pisces: The Fish/Duality, Unity, and Divination

Pronunciations: Pisces (PIE-seez); Piscium (PIE-see-em)

Visible Latitudes: 90° North to 65° South

Constellation Abbreviation: Psc

Bordering Constellations: Andromeda, Aquarius, Aries, Cetus, Pegasus

Description: The stars form a large V shape that leans to one side, with the bottom of the V pointing toward the southeast. The westernmost end of the shape culminates in a circle.

To Find: Locate the Great Square of Pegasus. Below it is the circle of stars that depict one of the fish. The other half of the V shape, closer to Andromeda, represents the other fish.

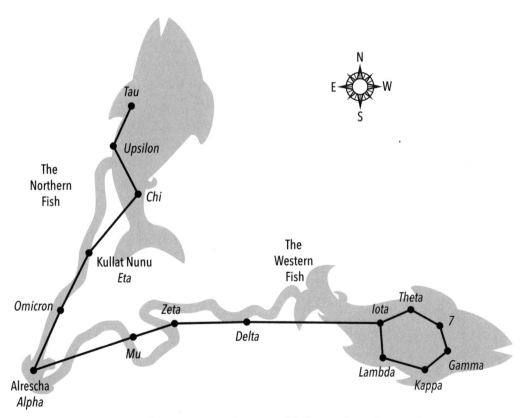

Figure 6.9. Pisces represents the power of the knot and creative energies.

According to modern astrology, Pisces is the last constellation of the zodiac; however, because of precession it now hosts the vernal equinox. In Latin, Pisces means "the fish." The circlet of stars represents the head of the Western Fish. The line of stars closer to Andromeda is known as the Northern Fish. Like the other marine-related constellations, Pisces is located in an area of the sky that the ancients called the Sea. The Babylonians, Persians, Greeks, and Turks regarded this constellation as two fish. The Arabs considered the Western Fish as such, but regarded the stars of the Northern Fish as part of the Andromeda constellation.

The Babylonians considered Pisces as a pair of fish joined by a cord, as did the Greeks and the Romans. The constellation is usually associated with the Roman myth of Venus and Cupid, who tied themselves together with a rope and shape-shifted into fish to escape the monster Typhon. For a time, the constellation was commonly known as Venus and Cupid. If this sounds familiar it is because Pan, in the story of Capricornus, jumped into a river and partially turned into a fish to escape the same monster.

In Greek myth, Aphrodite and her son Eros called to the water nymphs for help before jumping into a river. In one version of the story, two fish came to the rescue and carried Aphrodite and Eros to safety on their backs. In another version, mother and son were themselves transformed into fish.

In ancient Egypt, this constellation was a sign that springtime and the fishing season were about to commence. According to Ptolemy, the stars of Pisces were equated with Mercury and Saturn or Jupiter, depending on their location in the constellation. In modern astrology, Pisces is equated with Neptune. For medical purposes, Culpeper determined that Pisces influenced the feet, toes, and thalamus gland.

Notable Stars in Pisces

Official Designation: Alpha Piscium

Traditional Name: Alrescha

Pronunciation: ahl-RESH-ah

While Alrescha is only the third-brightest star in the constellation, it holds the important intersecting point of the two lines of stars at the bottom of the V shape. This white, double star marks the knot of the rope. Its name comes from Arabic and means "the cord." Ptolemy called it the Bright Star in the Knot. Alrescha is associated with acquiring knowledge and integrating ideas.

Official Designation: Eta Piscium

Traditional Name: Kullat Nunu

Pronunciation: KU-lot new-new

This yellow star is the brightest in the constellation. The name Kullat Nunu is believed to come from the Babylonian words for "cord" and "fish," respectively. It is usually depicted as part of the cord just below the Northern Fish.

Official Designation: Gamma Piscium

This yellow star is the second brightest in the constellation. It is part of the Circlet of Pisces asterism, which represents the Western Fish. This star does not have a traditional name.

Magical Interpretations and Uses for Pisces

In Egypt, Mesopotamia, and throughout the ancient world fish were a symbol of fecundity and the feminine principle. To the Greeks, a fish was a symbol of Aphrodite. The *Vesica Pisces*, "vessel of the fish," is the central almond shape created between the two fish in many depictions of them. It is a symbol that represents the fertile creative power of the vulva. The almond-shaped yoni symbol dates back to the Neolithic period and to the worship of the Great Mother Goddess. It remained a powerful symbol that survived as the *sheela-na-gig* portrayed on old churches in Ireland, which shows a goddess squatting to display her yoni.

Like a seed resting in the darkness of the womb or in the earth (as we head into winter), this state of incubation beckons to the future and what is to come. To foster your creative and procreative energies, gather fifteen pieces of bloodstone, coral, moonstone, or pearls. On the night of a new moon, hold them in your hands as you draw down the energy of this constellation. Release the energy into the stones as you say: *"Creative forces stirring in me; With aid from Pisces set these free. To nurture and grow as in a womb; Begin to draw life on this dark moon."* Say this three times and then lay out the configuration of the Pisces's star pattern wherever you do your creative work or in your bedroom if you are trying to conceive.

In legend, not only are the two fish of Pisces mentioned as joined by a cord, but a cord that is knotted. The alpha star, Alrescha, represents that knot. Around the world and throughout the centuries knotted cords were believed to hold mystical properties. They were used in medical practices, witch charms, and folk traditions. Sailors believed they could control the winds by knotting and unknotting special cords. A knot, of course, is the symbol of a bond, and a knotted cord used as part of your Samhain remembrance ritual is a symbol of the unbroken bonds with those you love. In preparation for setting up your sabbat altar, take a length of string (white if you are using a dark-colored altar cloth) and make fifteen knots with a little space in between each of them. Cut the knots apart and use them to lay out the Pisces star pattern. Also cut a two-foot length of string. During your ritual when you pause to remember your loved ones and speak the name of each person, make a knot in the two-foot length of string. When you have named all those you are commemorating, hold the string in your hands as you think of how each person has touched your life. When you are finished, lay the cord around the Pisces star pattern on your altar, and then place a piece of candy in the center. This will carry the symbolism of sweet memories as well as the intention to sweeten the paths of your loved ones in the otherworld.

Chapter Seven

THE WINTER QUARTER OF DECEMBER, JANUARY, FEBRUARY

Summer is a fond memory and autumn's splendor has withered and faded. As we enter the final night of the year we are compelled to turn inward for a sense of balance that will hold us and carry us through the winter. It is a time for working with our inner power to develop our magical skills as well as a sense of who we are and our place in the web of existence.

The fields are empty and the sky is a monotone gray, but the magic of this season can be seen in the holly. The bright green and red of leaves and berries hold the promise of ongoing life. Like holly, evergreen trees were considered sacred because they didn't seem to die each year, and so they came to represent the eternal aspect of the Goddess. Holly is a symbol of the winter solstice, a time of joyous transformation as we await the return of the sun/son.

As the final night of the year draws to a close, step outside and reflect on life under the stars. This is the traditional time to take stock of our personal journeys during the previous twelve months and to prepare for the year ahead knowing where we have been and where we intend to travel. Winter storms may howl, but appreciating the beauty of this fierce side of nature means that we can flow with and enjoy the spiral of life energy rather than struggle against it. Just as a storm can be most fierce before it ends, winter can be most brutal just before spring. However, the increasing length of daylight gives us hope. Even with the icy grip still upon us, we can feel the world beginning to awake.

Although stargazing may be a rather chilly prospect for many of us, the crisp winter air seems to make the stars brighter. This time of year has been called the Capella Quarter, named for the alpha star in the Auriga constellation. In addition to Auriga the Charioteer, we will find a set of twins, a hero, a hunter, and seven sisters. Also, the animal kingdom is well represented with two dogs, a hare, a unicorn, a bull, and a crab with a beehive. There is also a river and a region called the Sea.

As in previous chapters, some of the bordering constellations noted here may fall within other seasons. The directions given to locate constellations and stars assumes that the reader is facing south. Don't forget to look at chapter 8 as some of the southern constellations listed only in that chapter may be visible to you.

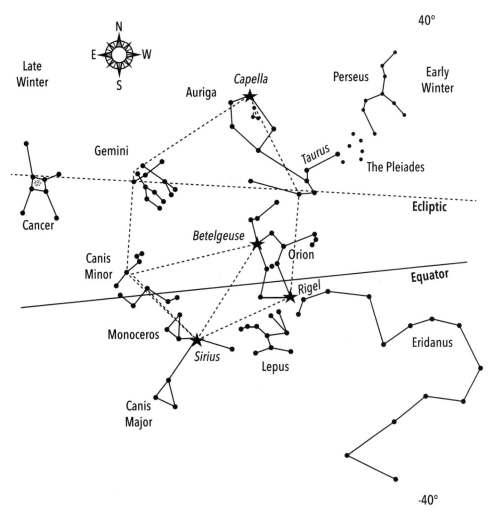

Figure 7.1. The winter night sky. The dotted shapes show the Winter Triangle and Winter Hexagon asterisms. Four of the brightest stars are also noted.

The Winter Constellations

Auriga: The Charioteer/Wild and Wise

Pronunciations: Auriga (oh-RYE-gah); Aurigae (oh-RYE-gay)

Visible Latitudes: 90° North to 40° South

Constellation Abbreviation: Aur

Bordering Constellations: Gemini, Perseus, Taurus

Description: An uneven pentagon shape with a faint triangle of stars.

To Find: Locate Orion's Belt, and then draw an imaginary line north to find the bright star Capella in Auriga. Capella marks the top anchor point of the Winter Hexagon asterism. The star at the bottom of Auriga's pentagon shape is actually the beta star in Taurus.

Auriga is usually depicted either as a man sitting on the Milky Way or as a charioteer holding a goat and its young. There is no myth that explains why this figure is carrying goats. The faint triangle of stars in the constellation are called the Kids, which at one time were considered a separate constellation. While the Greek and Arab names for Auriga translate as "the holder of the reins," Assyrian tablets represented this constellation more like a chariot than a charioteer.

Auriga is most often identified with Erichthonius, a mythical king of Athens who was raised by the goddess Athena. According to legend, he was the first to tame and harness horses for pulling chariots. His abilities impressed Zeus so much that the god placed him among the stars as a tribute. Auriga is also said to represent Hephaestus, the god of crafts and fire, who was lame and unable to get around easily. In order to travel where and when he wanted, he invented the chariot.

According to Ptolemy the stars of Auriga are equated with Mars and Mercury.

Notable Stars in Auriga

Official Designation: Alpha Aurigae

Traditional Name: Capella

Pronunciation: kah-PELL-ah

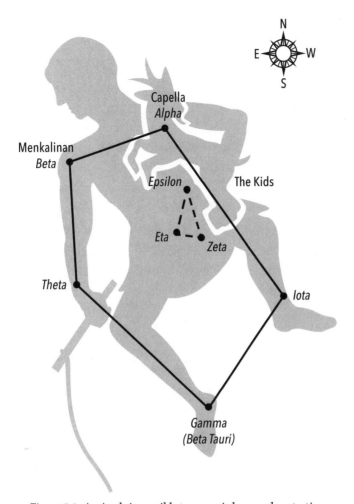

Figure 7.2. Auriga brings wild storms, wisdom, and protection.

The sixth brightest star in the sky, Capella is actually a pair of binary stars. The Alpha-1 stars are yellow and Alpha-2 are red. In depictions of the charioteer, Capella marks the heart of the goat he is holding on his left shoulder. The name *Capella* is Latin, meaning "female goat." To ancient Romans, this celestial animal represented the she-goat that suckled the infant Jupiter. To Babylonians, Capella was the star of Marduk, their powerful creator god. The Arabs called it the Guardian of the Pleiades. The energy of this star is equated with Mars and the moon. In addition, it was one of Agrippa's fifteen important fixed stars.

Official Designation: Beta Aurigae

Traditional Name: Menkalinan

Pronunciation: men-KAH-lee-nan

Derived from Arabic, this star's traditional name means "the shoulder of the rein-holder." Menkalinan is a blue-white binary star.

Official Designation: Gamma Aurigae / Beta Tauri

Like Alpheratz, the alpha star in Andromeda that is part of the Great Square of Pegasus, this star has a dual function. It marks the charioteer's right foot in Auriga and the tip of one of Taurus the Bull's horns. While it is now officially in the Taurus region of the sky, this blue star holds designations in both constellations.

Magical Interpretations and Uses for Auriga

Legends from many cultures tell about gods and fairies traveling in chariots that can traverse the land, sea, and sky. I think of Auriga as the Norse god Thor. He is a god of sky and storms, and the clattering wheels of his chariot were said to create thunder. Although we mainly associate thunder with summer storms, thunder-snow (thunder during a snowstorm) is not all that unusual, at least where I live in northern New England. Perhaps a more interesting association between this constellation and Thor is that his chariot was said to be pulled by two goats.

Even though he is formidable, Thor represents good against evil and light against dark. This was significant for the dark months of winter at a time when homes were lit by candlelight. This mighty god maintains the hope of the returning sun that winter solstice brings. In addition to being a sky god, Thor is a protector of the common people and offers protection against winter storms. If you live in a place that gets a lot of snow, you may want to initiate this at the beginning of winter before the white stuff piles up. Draw the Auriga star pattern on four stones and place them at each corner of your house. Use the following incantation as you put them in place: *"As winter storms soon will howl; Let protection begin now. No matter what comes, this house stands strong; And keeps us safe all winter long."*

Before storms strike, have enough snowflake obsidian on hand to lay out Auriga's star pattern. The white snowflake pattern on a black background makes this gemstone symbolic of winter storms. Because snowflake obsidian is also associated with protection, it will help strengthen the energy of your intention. When a storm begins, lay out the Auriga star pattern on your altar and then draw down the energy of the constellation. When you feel it increasing, hold your hands over the pieces of obsidian and then visualize the energy flowing out to each corner of your house where you have placed the Auriga star configuration. Continue working with the energy until you can visualize it rising up and around your house, forming a magical shield over the roof. Hold the vision as you say: *"Power of Auriga awake and begin; Rise and shelter all within. Wind may howl and snow may blow; Keep us safe—make it so."*

Let's look at another side of winter and the charioteer. Snowfall can bring a hush that is deepened by tightly closed windows. This quiet world helps us move inward and have time with our thoughts. In Hindu mythology, a charioteer functioned more as a guide than a servant, and in Celtic lore he had high status as a trusted advisor. Combining this aspect of a charioteer with the depiction of Auriga as a wise man sitting on the Milky Way emphasizes this constellation as the keeper and purveyor of knowledge. This side of Auriga represents any number of gods associated with wisdom, so a quiet winter's night is a good time to pursue knowledge. It could take the form of reading and studying, working on divination or psychic skills, or spending time in meditation to access the wisdom within. Set aside a time and place for this and make it special "me" time that enriches your soul. As an aid, draw the Auriga star pattern on a piece of paper that becomes your bookmark, a mat for laying out divination tools, or something to hold while meditating. Each time before you begin your session, draw down the energy of Auriga to aid you in your practice.

Cancer: The Crab / Between the Worlds

Pronunciations: Cancer (KAN-sir); Cancri (KAN-kry)

Visible Latitudes: 90° North to 60° South

Constellation Abbreviation: Cnc

Bordering Constellations: Canis Minor, Gemini, Hydra, Leo

Description: An uneven square at the center with spokes protruding at various angles.

To Find: From the bright star Capella in Auriga, draw an imaginary line southeast toward Castor and Pollux, the heads of the Gemini twins. Continue that line and the next constellation is Cancer. It takes a dark sky to see this constellation.

Cancer is Latin for "crab"; however, instead of a crab, it represented a lobster or crayfish to medieval astronomers in Europe. Despite the fact that it is rather inconspicuous, this constellation was familiar to ancient stargazers. It was occasionally referred to as a "dark sign" because its stars are so faint. While the Babylonians considered Cancer a tortoise, the Egyptians regarded it as a scarab beetle and as such, a symbol of immortality. The Arabs considered Cancer as part of the Leo constellation and said it represented the mouth or muzzle of the lion.

In Greek and Roman mythology, Cancer represented the crab that Hera sent to distract Hercules while he was fighting Hydra the Water Snake. According to one version of the story, Hercules kicked the crab all the way to the stars when it tried to pinch him. In another version, Hercules crushed the crab and Hera placed it in the heavens for its efforts. However, she placed the crab in an area with no bright stars because it had not successfully fulfilled its task.

Similar to the southern Tropic of Capricorn, the Tropic of Cancer is an imaginary line north of the equator that marks the northernmost point at which the sun appears overhead at noon on the summer solstice. When this was noted two thousand years ago, Cancer was in position as a backdrop to the sun and the line was named for this constellation. Because modern astrology uses the traditional dates for the zodiac, Cancer is still the sign that begins summer; however, Gemini is the actual constellation in position on the other side of the sun on the solstice.

The Beehive is an open cluster of stars within the constellation Cancer. Since ancient times, the Beehive has been described as a mist or a cloud and it is one of the seven nebulae (clouds) noted by Ptolemy. Also known as Praesepe, the Beehive is a Messier object. The Beehive has been regarded as a cradle of life, which may come from the Chaldean belief that souls passed through a gateway located in the constellation of Cancer as they descended to earth. The Chaldeans called the position of Cancer the Gate of Men.

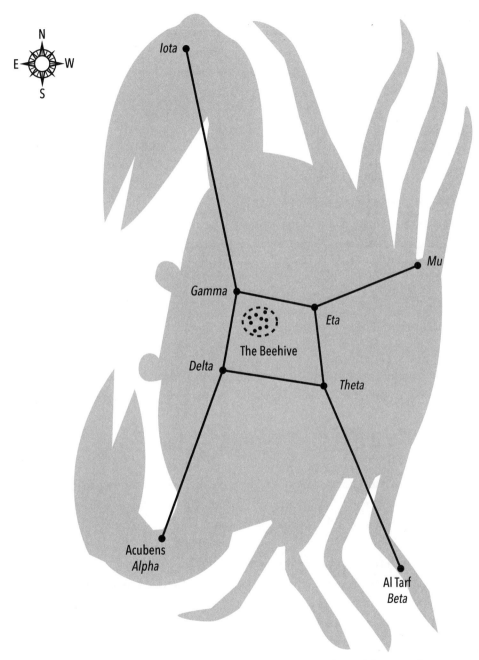

Figure 7.3. Use Cancer the Crab for astral travel and psychic and dream work, and use the Beehive for prosperity.

According to Ptolemy, most of the stars were equated with Mercury and Mars and those in the claws with Saturn and Mercury. In modern astrology, Cancer is equated with the moon. Culpeper determined that Cancer influences the breast, stomach, esophagus, and left side of the body. This constellation is associated with perseverance and life-force energy.

Notable Stars in Cancer

Official Designation: Alpha Cancri

Traditional Names: Acubens; Sertan

Pronunciations: ACK-you-benz; sir-TAN

Acubens is a white binary star that marks one of the crab's claws. The name is Arabic and means "the claw." Also from Arabic, *Sertan* means "the crab" and was the name for the entire constellation. Ptolemy equated this star with Saturn and Mercury.

Official Designation: Beta Cancri

Traditional Name: Al Tarf

Pronunciation: al TARF

Al Tarf is an orange binary star and the brightest in the constellation. Its name comes from Arabic and means "the end," in reference to this star marking one of the crab's back legs.

Official Designation: 44M Cancri

Traditional Name: Beehive

The Beehive looks like a fuzzy, oval glow. There are at least one thousand stars in the Beehive cluster, none of which have names or individual designations. Their colors run the gamut from red to orange, yellow, white, and blue.

Magical Interpretations and Uses for Cancer

As an inhabitant of the shoreline, the crab is a creature that is at home between the worlds because it dwells both in the water and on the land. As a result, this constellation can aid us in astral travel, in psychic and dream work, and for reaching into the imagination to stoke creativity. These endeavors require us to step outside of our mundane mindsets.

Moving sideways like a crab is a good analogy as we sidestep our daily routines to tap into the subconscious.

To do this, use water from the ocean if you can or create some seawater yourself. For this you will need a cup of fresh spring water and a tablespoon of sea salt. Boil the water, add the salt, and stir until it dissolves. After it cools, pour the water into a small bowl and place it on your altar. If you happen to have a little beach sand, spread it evenly on your altar or a plate and use your finger to make indentations to create Cancer's star pattern. If you don't have sand, use small seashells, or anything that you feel is appropriate.

Draw down the energy of Cancer as you focus your attention on your three celestial chakras. As you do this, dip a finger in the saltwater and touch it to your third eye chakra, located slightly above and between your eyebrows. Draw the energy from the celestial chakras to your third eye chakra. Briefly review the activity that you are planning for the evening. Later when you are finished with your activity, be sure to ground your energy and thank Cancer for support.

Honey was a valuable commodity in ancient times. It was used as food, medicine, and as an offering to deities. Bees were regarded as harbingers of wealth, and keeping bees symbolized abundance. With this association in mind, we can use the energy of the Beehive to boost the power of prosperity spells. Arrange a cluster of tea light candles on your altar to represent the Beehive. Place a small bowl or saucer with a spoonful or two of honey in front of the candles. Draw down the energy of the stars as you hold your hands above the bowl. Release it into the honey as you visualize prosperity coming to you, and then say: *"Honey from a bee and honey from a star; Bring prosperity from near and far."*

Dip a finger into the honey and then put it in your mouth as you visualize prosperity coming into your life. Then say: *"Stars of yellow, white, orange, red, and blue; Hear my wish and make it true."* Take more honey as you continue your visualization. Repeat this two more times. Hold the images and ideas of prosperity in your mind another moment or two, and then let them fade. Sit quietly in the light of the candles as you ground your energy. Blow out the candles, go outside, and look at the stars as you say: *"Star light, star bright, I call the Beehive Cluster tonight. This spell I send with all my might, may the spark of prosperity now ignite."*

Canis Major and Canis Minor: The Great Dog and the Little Dog/
   Guardians and Guides

Pronunciations: Canis Major (KAY-niss MAY-jer); Canis Majoris (KAY-niss mah-JOR-iss)

Visible Latitudes: 60° North to 90° South

Constellation Abbreviation: CMa

Bordering Constellations: Lepus, Monoceros

Description: To the left of and below the bright star Sirius, there are two triangle shapes. One marks the dog's head; the other its hindquarters.

To Find: Locate Orion and draw an imaginary line through his belt to the southeast. This will bring you to the star Sirius, which marks the southern anchor point for the Winter Triangle asterism.

Pronunciations: Canis Minor (KAY-niss MY-ner); Canis Minoris (KAY-niss my-NOR-iss)

Visible Latitudes: 90° North to 75° South

Constellation Abbreviation: CMi

Bordering Constellations: Cancer, Gemini, Hydra, Monoceros

Description: A straight line with a zigzag at one end.

To Find: From Sirius in Canis Major, draw an imaginary line to the northeast to the next bright star, Procyon, the little dog's alpha star. Procyon marks the eastern anchor point for the Winter Triangle asterism.

According to Greek myth these are the hunting dogs of Orion. Canis Major means "the great [big] dog" in Latin. Canis Major was sometimes known as the Dog with the Blazing Face because the star figure was often shown with Sirius, the brightest star in the sky, in its mouth. Modern depictions place Sirius closer to the dog's heart. Because of Sirius, Canis Major was one of the most important constellations in ancient times. Only the moon, Venus, Jupiter, and Mars are brighter than Sirius.

In Greek mythology, Canis Major is associated with Laelaps, the fastest dog in the world that could catch anything it pursued. Ownership of the dog differs in the numerous versions of the story. In one version, it was a gift from Artemis to Procris, the daughter of the king of Athens, and in another it was Amphitryon of Troezen. Whomever the owner,

Laelaps met his match with a fox that could not be caught. Zeus eventually intervened and put a stop to the endless chase by turning both animals to stone. He then placed them in the night sky.

Ptolemy equated all but the star Sirius with Venus.

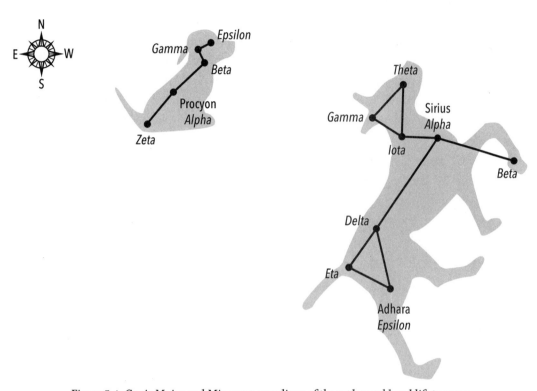

Figure 7.4. Canis Major and Minor are guardians of the underworld and life to come.

The name Canis Minor means "lesser [small] dog" in Latin. Informally it has been called the Puppy. According to one legend, it represented the dog Maera whose owner was Icarius the winemaker (not to be confused with Icarus who flew too close to the sun). In grief at finding his master dead, Maera threw himself in a river. It is unknown whether this story has any connection to Canis Minor having been known as the Water Dog in the Euphrates region. The Chinese also associated this constellation with water. In another myth, Canis Minor was identified as the Teumessian Vixen, the fox that the dog Laelaps (Canis Major) could not run down.

According to Ptolemy the stars of Canis Minor are equated with Mercury and Mars.

Notable Stars in Canis Major

Official Designation: Alpha Canis Majoris

Traditional Name: Sirius

Pronunciation: SEER-ee-us

Sirius is a white binary star and the brightest in the sky. The smaller component has been referred to as the Pup. The name Sirius comes from Greek and means "scorching" or "searing." In ancient times, this star rose just before the sun during the hottest period of the year, which is the source of the phrase "the dog days of summer." In Egypt, the rising of Sirius marked the annual flooding of the Nile, an occasion linked with the return from the dead of the god Osiris. The Egyptians called this star Sirius Isis, while the Greeks associated it with Pan.

Ptolemy equated the energy of Sirius with Jupiter and Mars. In medieval medicine, this star was believed to provide a cure for dropsy and to aid against the plague. Sirius was one of Agrippa's fifteen fixed stars. In both the Winter Triangle and the Winter Hexagon asterisms, Sirius is the southernmost anchor point.

Official Designation: Epsilon Canis Majoris

Traditional Name: Adhara

Pronunciation: ahd-HAR-ah

This blue-white star is the second brightest in the constellation. The name Adhara comes from Arabic and means "virgin." Along with Adhara, the delta and eta stars form a triangle on the dog's rear flank and were collectively known as the Virgins for reasons that are no longer known.

The Notable Star in Canis Minor

Official Designation: Alpha Canis Minoris

Traditional Name: Procyon

Pronunciation: PRO-see-on

This white binary star is the seventh brightest in the sky. Its name is derived from Greek and means "before the dog." It was also occasionally called Antecanis, which means the same thing in Latin. This star is one of the anchor points in the Winter Hexagon and the easternmost anchor in the Winter Triangle. Procyon was one of the important fifteen fixed stars.

Magical Interpretations and Uses for Canis Major and Minor

These celestial dogs of winter fulfill their dual role with the springtime dogs of Canes Venatici. Canes Venatici is associated with life-force energy; Canis Major and Minor with death. As a guardian of the underworld, we may think of a dog as protecting the dead; however, it is more likely that it is guarding the new life to come. Throughout time and around the world, dogs have served as guardians. The Babylonians considered even the image of a dog to be magically protective. Howling at the moon and growling at nothing has earned them the reputation of possessing supernatural powers and being able to see what people cannot.

To call on these powers as guardian for your home, buy a new welcome mat for your front door. Before putting it in place, consecrate it with sandalwood oil and draw the star patterns of Canis Major and Minor on the underside of the mat as you say: *"Big dog, little dog, bring your powers here; Ward off danger, chase away fear. Bark loud to warn of trouble; Bark soft to share in play. Canis Major, Canis Minor, help me every day."*

Dogs held a great deal of significance for Celtic people, so much so that author and scholar Miranda Green categorized them together with the horse and bull as domestic animals with sacred significance. Whether god or human, hunters take dogs along to serve as guides, guardians, and fetches. In this way canines aid in bringing food to the table and earn their place by the hearth. Of course, they have made their way into our hearts as well. With Canis Major and Minor in the night sky, winter is a good time to honor pets that have passed on to the next world. Place a picture of your pet on your altar or somewhere special. Cut out a white star shape, or place a white candle in a star-shaped candleholder next to the picture as you say: *"Sirius, Sirius, most famous dog of all; On your strength and wisdom I call. Hold and protect my beloved friend; Until such a day when I see him/her again."*

Eridanus: The River/Flowing Sacred Waters

Pronunciations: Eridanus (eh-RIHD-uh-nuss); Eridani (eg-IHD-uh-nee)

Visible Latitudes: 32° North to 90° South

Constellation Abbreviation: Eri

Bordering Constellations: Cetus, Hydrus, Lepus, Orion, Phoenix, Taurus

Description: A long meandering trail of stars that runs from the Northern to Southern Hemisphere.

To Find: Locate Orion's Belt, and follow an imaginary line southwest to the bright star Rigel that marks Orion's left foot. Cursa, the beta star in Eridanus, is to the west of Rigel. Follow the trail of stars west and then south. If you are located far enough south you can see the river meander east, and then south again.

Eridanus is a southern constellation of which only the top portion can be viewed by most people in the Northern Hemisphere. It is believed to be of Euphratean origin because of its depiction on cylinder seals from that region. On later star maps, it was often depicted as a river flowing from the waters poured by Aquarius. Eridanus has represented the river of life as well as a river leading to the otherworld.

Egyptian and Babylonian astronomers associated this constellation with the Nile and Euphrates Rivers, respectively. In Britain it was considered to be the River Avon. Greek poet Aratus referred to it as Eridanus, while others called it Potamos, which simply means "the river." Even though the Greeks generally associated it with the River Phasis, which was near the place where Jason and the Argonauts found the golden fleece, Homer regarded it as the ocean stream that encircled the world. Eridanus is most often regarded as Italy's River Po.

This constellation is associated with the story of Phaeton, the mortal son of Helios, and his friend Cycnus. After much wrangling with his father, Phaeton was allowed to drive the chariot of the sun. Helios warned him to stay on the circle of the zodiac and although Phaeton tried, he lost control of the chariot. The boys were thrown from the vehicle and fell to Earth. Cycnus, who became Cygnus the Swan, survived the fall, but Phaeton plunged into the River Eridanus and was lost. The Heliades, daughters of Helios, grieved over the loss of their brother "until they turned into amber-dropping poplar trees alongside the river." [30] This story, plus the fact that the River Po was an important trade route for transporting amber from the Baltic to Europe, strengthens its link to this constellation.

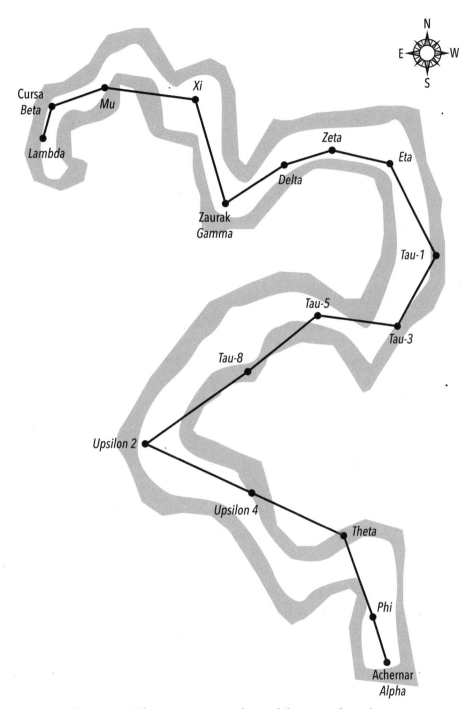

Figure 7.5. Eridanus represents wisdom and the power of sacred water.

Except for the star Achernar, Ptolemy equated this constellation with Saturn. Eridanus is associated with changes and authority.

Notable Stars in Eridanus

Official Designation: Alpha Eridani

Traditional Name: Achernar

Pronunciation: ACK-er-nar

Achernar is a binary star consisting of a blue primary with a white companion. Its name is Latin and means "the end of the river," which describes its location at the southern end of the constellation. It is not visible in the Northern Hemisphere. Ptolemy equated the energy of Achernar with Jupiter.

Official Designation: Beta Eridani

Traditional Name: Cursa

Pronunciation: COOR-sah

Cursa is a white star near the northern beginning of the river. Its name comes from an Arabic phrase meaning "the footstool of the central one," the name of a small (now extinct) constellation of which it was a part. The "central one" refers to Orion. Cursa is the brightest star in the northern part of the constellation.

Official Designation: Gamma Eridani

Traditional Name: Zaurak

Pronunciation: ZAW-rahk

The name of this yellow star is derived from Arabic and means "the bright star of the boat." Historians believe that at one time several stars may have represented a boat on the river Eridanus.

Magical Interpretations and Uses for Eridanus

The term "life-giving" does not only refer to giving birth, it also means to sustain life and provide nourishment. Recognizing that water ensures life, health, and abundance, people in ancient times associated these blessings and the power of water with deities. The water of springs and rivers seemed to flow from Gaia herself and was regarded as especially blessed. To the Celts, water was the boundary between our world and the otherworld.

They also considered it a boundary between the earth and the sky, and thus, it was a magical in-between place.

In many cultures, rivers were thought to have special powers or were used for worship. As a result, we find many goddesses associated with rivers. Some of these associations include Boann and Siann with the Rivers Boyne and Shannon, respectively, in Ireland; Danu with the Danube in Germany and Central Europe; Sequana with the Seine in France; Ganga with the Ganges of India; Oshun with the Osun in Nigeria; and Anuket with the Nile of Egypt.

The goddesses and their rivers represent power, wisdom, and transformation. We can call on the energy of river goddesses with the help of the celestial river Eridanus for ritual, spellwork, or meditation. To begin, place a blue cloth on your altar to represent both water and the sky. Lay out the Eridanus star pattern with silvery glitter/confetti or whatever you feel is appropriate to represent the stars. After drawing down the energy of this constellation, stand in front of your altar for a moment. Slowly raise your arms out to the sides as you chant the word "awen" (pronounced AH-OO-EN) three times. I think of this as the Celtic version of Om. *Awen* is a Welsh word that is usually translated as meaning "inspiration." While it is generally associated with bardic poets, it refers to the type of inspiration that is deeply spiritual and transformative, providing access to the source from which all creative possibilities flow. After chanting, slowly lower your arms and take a few moments to feel the power of Eridanus and the river goddesses flow over and through you before moving on to ritual, meditation, or other practice.

As a goddess of fire, Brigid's sabbat is Imbolc, when we celebrate the strengthening power of the sun. However, she also presides over sacred waters as the numerous springs and wells dedicated to her throughout Ireland attest. Fire and water are elements of purification and healing, and both stimulate new growth. An Imbolc ritual wouldn't seem complete without a lot of candles, and balancing the fire energy with water can actually strengthen the ritual. To incorporate Eridanus into your ritual, you will need a wide, shallow bowl and tea light candles. If possible, melt some snow or ice to fill the bowl. Before floating the candles on the water, draw down the energy of Eridanus. With your hands above the bowl, release the star energy to charge the water as you say: *"Eridanus, may your starry waters shine blessings on this altar. So mote it be."*

Gemini: The Twins/Power Doubled

Pronunciations: Gemini (JEM-in-eye); Geminorum (JEM-eh-NOR-um)

Visible Latitudes: 90° North to 60° South

Constellation Abbreviation: Gem

Bordering Constellations: Auriga, Cancer, Canis Minor, Monoceros, Orion, Taurus

Description: A pair of stick figures side by side.

To Find: Locate Orion's Belt and then draw an imaginary line toward the northeast. Look
for the two bright stars, Castor and Pollux. These stars represent the heads of the twins,
while fainter stars stretching to the southwest form two stick-figure bodies.

The name *Gemini* means "the twins" in Latin. The ancient Greeks regarded this constellation as Castor and Pollux, sons of the Spartan queen Leda. The Romans considered the twins as Romulus and Remus, the brothers who founded Rome.

In Greek mythology, Castor and Pollux were also known as the Dioscuri, which means "the sons of Zeus." However, in many versions of their story the twins are not twins with each other, nor is Castor a son of Zeus. Polydeuces (Pollux) and Helen (later to become Helen of Troy) were twins from the union of Leda and Zeus. Queen Leda was raped by the lusty god who visited her in the form of a swan. That myth is associated with the constellation Cygnus. King Tyndareus of Sparta, Leda's husband, fathered Castor and his sister Clytaemestra. Castor was mortal; Polydeuces was immortal.

Castor and Polydeuces grew up together and further legends include them with the Argonauts' expedition to attain the golden fleece. In addition, the twins were said to have been given the power to rescue shipwrecked sailors. Like most twins of legend, they represented a polarity of light (Castor) and dark (Polydeuces). This occurs in Celtic myth with the sons of Arianrhod; Lleu as the light and Dylan as the dark. It also holds true in Egyptian legend with Horus representing light and Seth, the dark.

According to Ptolemy, the stars of Gemini are equated with Mercury, Venus, or Saturn, depending on their location. Modern astrologers equate Gemini with Mercury. For medicinal purposes, Culpeper determined that Gemini influenced the lungs, shoulders, arms, hands, and sympathetic nervous system.

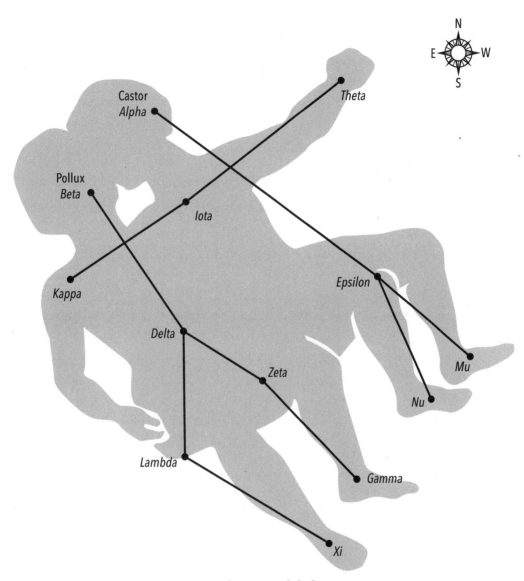

Figure 7.6. Gemini is a dynamic symbol of opposites.

Notable Stars in Gemini

Official Designation: Alpha Geminorum

Traditional Name: Castor

Pronunciation: KASS-ter

Despite its alpha designation, Castor is the second-brightest star in the constellation. It is a multiple star system composed of three spectroscopic binary stars, making it a sextuplet. How perfect that one of the mythological twins should have his name on a star system with three sets of twins. Both Alpha-1 and Alpha-2 sets of stars are white and the Alpha-3 stars are red. Ptolemy equated Castor with Mercury. This star is associated with success in legal matters.

Official Designation: Beta Geminorum

Traditional Name: Pollux

Pronunciation: POL-lucks

Pollux is an orange star and the brightest in the constellation. In Arabic it is known as the "Head of the Second Twin." Ptolemy equated this star with Mars. Pollux is associated with bravery and craftiness.

Both Castor and Pollux mark the twins' heads and serve as an anchor point in the Winter Hexagon asterism. Their traditional names are actually ancient Greek names. Ptolemy associated these two stars with Apollo and Hercules, who were half-brothers and both sons of Zeus.

Magical Interpretations and Uses for Gemini

The number two represents anything of a binary nature and serves as a dynamic symbol of opposites: light/dark, sun/moon, male/female, and good/evil. Because Gemini is the actual backdrop for the sun at the summer solstice and is visible in the night sky at the winter solstice, we can draw a parallel with the legend of the Oak and Holly Kings. These two sacred trees represent the dual aspect of nature. Oak rules the waxing half of the year beginning at Yule and the Holly King rules the waning half starting at summer solstice. To honor this cycle of change, mark a holly leaf with the Gemini star pattern, and draw down the power of this constellation to enhance the energy of your Yule altar or household decorations.

New Year's Eve is another occasion to call on the energy of the twins because it is a night of looking back at what has been and of looking forward to what will be. The energy of Gemini can aid in balancing this transition from old to new year. If you are celebrating at home and want to be discreet, decorate your table with star glitter/confetti. This can look festive with the stars sprinkled randomly across the tabletop. Before your guests arrive, choose a spot on the table that won't be disturbed where you can lay out the Gemini star pattern. As you do, say: *"From your place in the sky; Shine your light, Gemini. Illuminate the past, let us see it anew; Then on to the future and all we'll pursue."*

While the number two can represent division, it also represents the strength of unity. Because of this, the Gemini constellation can double the power of your spells. To do this, prepare two green candles by carving half of the Gemini star pattern into one and the other half into the second candle. Anoint the candles with bergamot, lavender, peppermint, or yarrow oil or a combination of these, and then set them side-by-side to form the constellation. Light the candles and draw down the energy of the constellation as you say: *"Candles burn, cauldron bubble; Castor and Pollux this spell to double. Carry my wishes, increase my will; Castor and Pollux this spell fulfill."* Proceed with your spellwork and when you are finished blow out the candles.

Lepus: The Hare/Magical Moon Energy
Pronunciations: Lepus (LEE-pus); Leporis (lee-POR-iss)
Visible Latitudes: 63° North to 90° South
Constellation Abbreviation: Lep
Bordering Constellations: Canis Major, Eridanus, Monoceros, Orion
Description: A stick figure animal with three stars forming a V shape for its ears.
To Find: Lepus is directly south of Orion and west of the bright star Sirius in Canis Major.

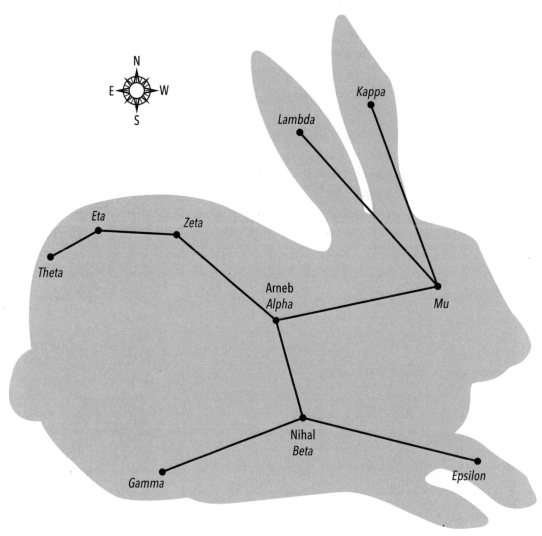

Figure 7.7. Lepus provides a boost of energy for divination and moon magic.

Lepus is a constellation of the Southern Hemisphere. Its name means "hare" in Latin. Lepus is usually portrayed as being chased by the hunter Orion and his dogs toward the River Eridanus. According to Greek legend, Hermes, the messenger of the gods, admired the swiftness of this animal and placed the hare among the stars as tribute. The Arabs called this constellation the "Throne of the Great One" due to its position underneath Orion. Because the Egyptians associated the Orion constellation with Osiris, they considered Lepus as the boat that carried his soul in the afterlife.

According to Ptolemy, the stars of Lepus are equated with Saturn and Mercury. In medieval medicine, the constellation was believed to guard against madness. The constellation is associated with abundance, fertility, and a quick wit.

Notable Stars in Lepus

Official Designation: Alpha Leporis

Traditional Name: Arneb

Pronunciation: ARE-neb

The traditional name of this yellow-white star comes from Arabic and means "the hare."

Official Designation: Beta Leporis

Traditional Name: Nihal

Pronunciation: nih-HALL

The traditional name of this yellow binary star means "quenching their thirst." This is a holdover from the time when it was part of an older constellation that the Arabs regarded as camels drinking from the nearby celestial river of Eridanus.

Magical Interpretations and Uses for Lepus

Hares are different from rabbits. They tend to be larger, with longer ears and bigger feet. Hares live above ground, while rabbits burrow. Unlike rabbits, hares have not been domesticated. They represent the abundance and wildness of the forest and symbolize fertility and swiftness. The Celts considered the hare a powerful spirit animal and associated it with their war goddess Andraste. Queen Boudicca was said to have invoked the power of Andraste to divine the outcome of a battle by observing how a hare ran when it was released from a cage. Likewise, we can call on the energy of Lepus for aid in our divination practices.

To boost the energy of divination tools, create a special cloth in which to wrap them when not in use. Buy a gray or brown piece of fabric large enough to wrap your tools. If necessary, stitch a hem around the edges to prevent the fabric from unraveling. Use a silver or white gel pen or silver or white thread to create the Lepus star pattern in the center of the fabric. When it is finished, wrap your tools in the cloth and draw down the energy of the constellation to charge them and the cloth.

The hare is associated with the moon, and in many cultures it is a hare, not a man's face, that people see on the lunar surface. Like the moon, the hare is a shape-shifter representing transformation and hidden knowledge. The energy of Lepus can enhance esbat rituals and boost moon magic. It also aids in working with the energy of the new moon. To do this, infuse a little bit of olive oil with ginger, juniper, lemon, or rosemary for use when preparing altar candles. These plants are associated with lunar energy. Place a few short stems and leaves of juniper or rosemary, pieces of ginger root, or lemon rind in a small jar and then pour in enough oil to cover them. Set the jar in a cool, dark place for at least a week, and then prepare a white or silver candle. Because it will be used to draw down the energy of the stars, put a little oil on the tip of your finger and run it from top to bottom four times around the candle to represent the cardinal directions.

After lighting the candle, lay out the Lepus star pattern on your altar with moonstone, selenite, or turquoise gemstones. Draw down the energy of the constellation to charge the candle and the gemstones as you say: *"Lepus, Lepus, magical hare; Mysteries of the moon with me share."* Gaze at the candle and visualize a hare. Reach out with your energy to sense its wildness and wisdom. Slowly bring your attention back to the candle flame and then extinguish it. On the next full moon, go outside and look at the moon to see if you can discern the hare. Once you do, you will have a special bond with this animal and the Lepus constellation for working moon magic.

Monoceros: The Unicorn/Power of the Crone

Pronunciations: Monoceros (mon-OSS-sir-us); Monocerotis (mon-oss-er-OH-tiss)

Visible Latitudes: 75° North to 90° South

Constellation Abbreviation: Mon

Bordering Constellations: Canis Major, Canis Minor, Gemini, Hydra, Lepus, Orion

Description: This constellation looks like a wide and somewhat flat letter *W*.

To Find: Monoceros is a relatively faint constellation to the northeast of the bright star Sirius in Canis Major. It is located between the two dogs Canis Major and Minor and east of Orion.

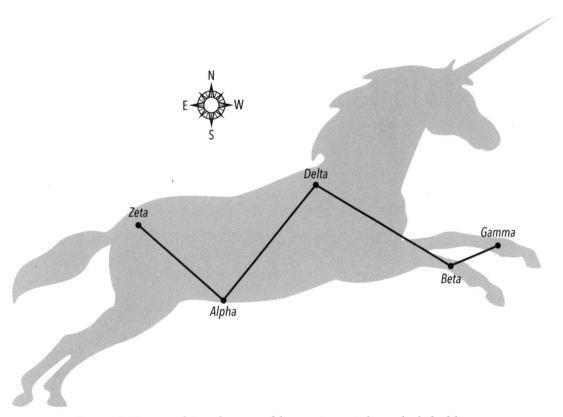

Figure 7.8. Monoceros brings the power of the crone to magic during the dark of the year.

Located on the celestial equator, this constellation's name is Greek for "unicorn," and it represents the mythical horselike creature. Like most modern constellations, it was created to fill in an area of sky and is not associated with any particular myth. Because of its location below Canis Minor, Monoceros is sometimes depicted like a circus pony with a dog on its back.

The unicorn constellation made its debut in 1612 on a globe created by the Dutch cartographer Petrus Plancius (1552–1622). He called it Monoceros Unicornis, using both the Greek and Latin words for unicorn. In 1624, German astronomer Jakob Bartsch (circa

1600–1633) included it in his star atlas, but he shortened the name to Unicornus. When it was made an official constellation in 1922, the Greek name was used.

Although this constellation is not linked with any particular myth, unicorns are legendary creatures that have appeared in folklore, art, medicine, literature, and magic. While their descriptions differ slightly, unicorns are almost universally known. Traces of their history can be found as far back as 3500 BCE in Mesopotamia.

Notable Stars in Monoceros

Official Designation: Alpha Monocerotis

Official Designation: Delta Monocerotis

Official Designation: Gamma Monocerotis

Despite its rank as third, Gamma Monocerotis is the second-brightest star in the constellation. The alpha star marks the belly of the unicorn, and the gamma star a forefoot. Both the alpha and gamma stars are orange. Delta is a triple white star and it is notable because its three components form a triangle, which is the shape of this Greek letter capitalized. Discovered in 1781 by Sir William Herschel, it was described as one of the most beautiful sights in the sky. The delta star marks the unicorn's shoulder.

Magical Interpretations and Uses for Monoceros

Although the unicorn has been well known for thousands of years, its popularity reached a zenith during the Middle Ages and Renaissance in Europe when it was widely used in literature, heraldry, art, and medicine. Finely ground powder from a unicorn horn was believed to be a potent aphrodisiac, however, according to many myths only a virgin was capable of capturing the animal. A captive unicorn was said to bring luck and love to its keeper. Such a creature is the subject of the famous, medieval unicorn tapestries. The mystery of the woman featured with the unicorn has fascinated writers, artists, and poets for centuries. I was also captivated by these tapestries, and as a teenager I would take the long train ride to the Cloisters Museum in upper Manhattan to see them as often as I could.

The unicorn has served as a symbol of beauty, enchantment, freedom, love, luck, spiritual revelation, transformation, and wisdom. In earlier times, it was considered a powerful and dangerous animal, although today it is ascribed a more gentle nature and associated with virgins. However, I agree with D. J. Conway's interpretation that instead of the maiden, the unicorn carries the power of the crone in that purity and transformation come from destroying the old to make way for the new.[31] There is no glitter or fluffy bunnies with the celestial unicorn because it represents the cycle of elemental power. The vitality of summer is dead and gone, but this is the time of incubation in the dark womb of winter. That is the wisdom and enchantment of this legendary beast and the power of the crone.

Call on this constellation to deepen spiritual meditations, to stoke the flame of creativity, and to initiate self-transformation. Carve the Monoceros star pattern into a white candle or lay out the pattern with pieces of white quartz. Draw down the energy of these stars and concentrate it in your third eye chakra. Hum the sound of the letter *M* for a moment or two and let it vibrate through this chakra. Visualize what you want to incubate and manifest into your life, and then allow the images to slowly fade. Give thanks to Monoceros and end the session.

Orion: The Hunter/Artemis the Huntress

Pronunciations: Orion (oh-RYE-un); Orionis (OH-rye-OH-niss)

Visible Latitudes: 85° North to 75° South

Constellation Abbreviation: Ori

Bordering Constellations: Eridanus, Gemini, Lepus, Monoceros, Taurus

Description: An uneven hourglass shape with three bright stars across the narrow middle.

To Find: The distinctive arrangement of three bright stars in a straight line marking Orion's Belt makes this constellation easy to find. Located southwest of the belt is the bright star Rigel, which represents a foot of the hunter.

Orion is one of the best-known constellations and contains two of the ten brightest stars in the sky. In addition to associating this constellation with the myths of Tammuz, the Babylonians called it the Heavenly Shepherd. The early Greeks equated the star figure with the beautiful Adonis and called it the Lord of the River Bank, in reference to the nearby Eridanus constellation. Later Greek legends tell of Orion bragging about his prowess and his ability to kill any wild beast on earth. This, of course, offended the goddess Gaia, mother of all animals. In retaliation, she sent one of her deadliest creatures, the scorpion, to kill him.

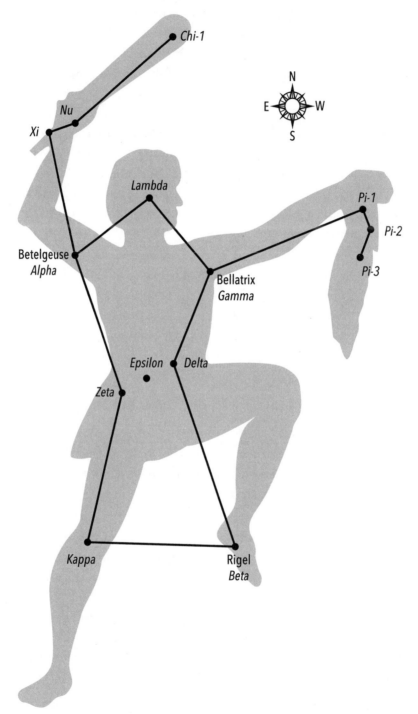

Figure 7.9. Orion leads the Wild Hunt of tempestuous winter storms.

There is another myth about his death that does not involve a scorpion. Who better to fall in love with the mighty hunter, Orion, than Artemis, the goddess of hunting. To prevent his sister from giving up her vow of chastity (of which she was fiercely protective), Apollo presented Artemis with an archery challenge. Not knowing that the target was Orion, she inadvertently killed him. In her sorrow, she placed him in the sky and gave him one of her dogs, Sirius, to keep him company.

The Celts of Ireland called this constellation the Armed King. The Syrians called it the Giant. In South Africa, the three stars of Orion's Belt are known as the Three Kings as well as the Three Sisters. The Egyptians associated Orion with Osiris, the god of death, afterlife, and rebirth. They believed that Osiris's soul rested in the constellation and regarded the Lepus constellation underneath Orion as the god's boat.

Except for Betelgeuse, Ptolemy equated the stars of this constellation with Jupiter and Saturn. Orion is associated with self-confidence, strength, and victory.

Notable Stars in Orion

Official Designation: Alpha Orionis

Traditional Name: Betelgeuse

Pronunciation: BEET-el-jooz

This famous red star is the second brightest in Orion. Located on the hunter's right shoulder, Betelgeuse is one of the largest stars known to astronomers. It is also one of the anchor points for the Winter Triangle asterism. The name Betelgeuse is the result of mistranslations from Arabic into medieval Latin. As mentioned in chapter 1, Arab astronomers saw this constellation as a female figure they called al-Jauza and often referred to it as the Central One. The Arabic name for Betelgeuse *yad al-Jauza* meant "the hand of al-Jauza." In medieval times when it was translated into Latin it became *bad al-Jauza* meaning "the armpit of al-Jauza." A later mistake during the Renaissance changed it to *bat al-Jauza* and it eventually became Betelgeuse. According to Ptolemy, this star is equated with Mars and Mercury.

Official Designation: Beta Orionis

Traditional Name: Rigel

Pronunciation: RYE-jel

Rigel is another red star and the brightest in this constellation. It also ranks as the sixth brightest star in the sky. Rigel marks the left foot of Orion. This star's traditional name comes from an Arabic phrase *rijl al-Jauza*, "the foot of al-Jauza." As mentioned in chapter 1, the word "rijl" evolved into "rigel" and this remained as the star's name. The alternate spelling "rigil" is used in the alpha star in the Centaurus constellation.

Official Designation: Gamma Orionis

Traditional Name: Bellatrix

Pronunciation: BEL-lah-tricks

Bellatrix is a blue star located on Orion's left shoulder. Its traditional name is Latin and means "the female warrior." Because of this, it has been nicknamed the Amazon Star.

Magical Interpretations and Uses for Orion

This constellation is associated with male hunter gods such as Herne, Cernunnos, Gwyn ap Nudd, Odin, Woden, and Nodens. Legends of the spectral Wild Hunt persist in Europe and tell of a cavalcade of hunters and dogs that thunder across the sky in a wild chase. There are many and varied interpretations of the Wild Hunt, one of which simply equates it with the wildness and unpredictability of nature. Additionally, from Greece to Persia and to India, Orion was regarded as a stormy constellation and so we can call on Orion to aid us in connecting with the power of tempestuous winter storms.

A winter storm can be used to boost a spell. Also, as an archer Orion can empower us to target and reach goals. Lay out the Orion star pattern on your altar with pieces of agate, beryl, or jet at the beginning of a storm. Draw down energy from the constellation and call on any of the hunter gods for protection. Dress appropriately and go outside, but only if you can do so safely. Otherwise, watch from a window where you can observe the weather and its effects. Reach out with your energy to feel the force of the storm, and then close your eyes and experience the raw wildness of nature. When the intensity of the energy feels at a crescendo, return to your altar and proceed with your spellwork, directing the strength of the storm toward your goal.

Because her story is intertwined with Orion's, I like to think of this constellation as Artemis herself. Besides, when the star figure is depicted with the hourglass shape it looks more feminine than masculine. This also fits with the constellation having represented a female figure known as al-Jauza.

Artemis is known as a goddess of the hunt, forests, and animals as well as a moon goddess. On the night of a full moon, lay out the star pattern on your altar with pieces of moonstone. Hold a small amount of dried mugwort in your hands. This herb's botanical name, *Artemisia vulgaris*, was derived from the goddess's name. The silvery color of its leaves gives the plant an ethereal, moonlike, and frosty appearance quite appropriate for winter.

Visualize coming into your own power, developing your independence, and goals you want to achieve. Do this as you draw down the energy of the constellation. Release the energy into the mugwort, and then burn it in your cauldron or other safe place. Raise your arms with your palms facing upward as you say: *"Mistress of the moon, who runs with the wild creatures of the forest, grant me your blessings so I may come into my power and attain what I seek."* When the mugwort ashes have cooled, take them outside and scatter them near a tree. If there is a breeze, release the ashes into the air and allow the wind to carry them away.

Perseus: The Hero/Returning God of Light

Pronunciations: Perseus (PURR-see-us); Persei (PURR-see-eye)

Visible Latitudes: 90° North to 35° South

Constellation Abbreviation: Per

Bordering Constellations: Andromeda, Aries, Auriga, Cassiopeia, Taurus

Description: A crooked, upside-down letter *Y*.

To Find: From Orion's Belt, draw an imaginary line north to Capella, the bright star in Auriga, the Charioteer. Perseus is west of Auriga.

Although Perseus is part of Andromeda's story, his constellation actually falls into the winter quadrant. In addition, we will see how he is equated with sun gods, which makes him more appropriate to the season of Yule. Also called the Champion, this constellation was known as Parash, "a horseman," to the ancient Hebrews.

In Greek mythology, Perseus was the hero who slew Medusa and hitched a ride on the winged horse Pegasus that was formed at her death. Finding himself in the right place at the right time, Perseus was able to rescue Andromeda from the sea monster.

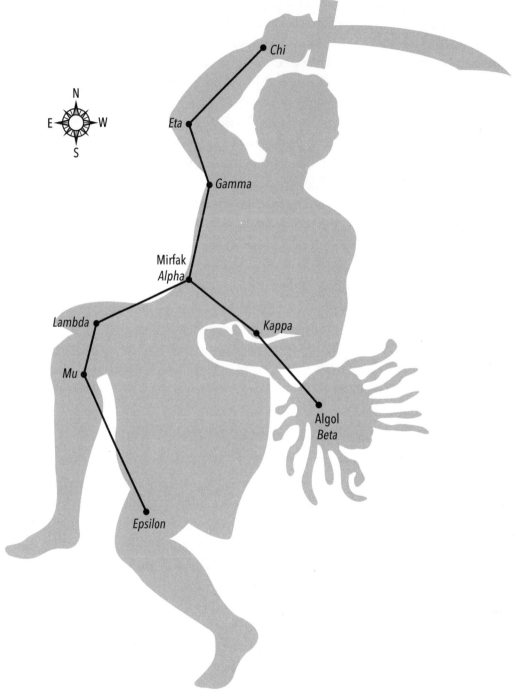

Figure 7.10. Perseus enhances Yule wishes and heralds the return of the sun/son.

The Greek Perseus has been equated with the solar deity Mithras of Persia. To the Iranians, this god was known as Mithra as well as Zarathustra, and in India he was Mitra. Central to a mystery religion that lasted for almost 500 years, the Roman Mithras was a fusion of the Greek Perseus and the Persian Mithras. It didn't take a great stretch of imagination to find similarities between gorgon slayer and bull slayer, especially when legends about both of them begin with their births taking place in a cave.

Perseus and the Celtic Lugh also have parallels in their stories. When a prophecy revealed to Perseus's grandfather that his grandson would kill him, he imprisoned his daughter in an attempt to prevent her from having a child. In Celtic legend, after Balor heard a prophecy that his grandson would kill him, he locked up his daughter Eithne. In both stories the daughters manage to get free and have sons, who through happenstance kill their grandfathers.

Ptolemy equated the stars of Perseus with Jupiter and Saturn. In medieval medicine, these stars were believed to free a person from the influences of witchcraft. This constellation is associated with protection and breaking hexes.

Notable Stars in Perseus

Official Designation: Alpha Persei

Traditional Names: Mirfak; Algenib

Pronunciations: MERE-fahk; al-JEN-nib

This yellow-white star is one of the brightest in the sky. Both of its traditional names come from Arabic. Mirfak means "elbow" and Algenib means "flank" or "side." If the second name is familiar, it is because the gamma star in Pegasus is also called Algenib.

Official Designation: Beta Persei

Traditional Name: Algol

Pronunciation: AL-gahl

The traditional name of this eclipsing binary star is derived from Arabic and means "the head of the demon." In many depictions of this constellation it represents Medusa's eye because the star seems to wink. When the dimmer companion star, Beta-2, passes in front of the bright Beta-1, the effect is a decrease in brilliance, which produces a winking effect. Algol is actually a triple star. Beta-1 is blue-white, Beta-2 is orange, and Beta-3 is white. The Hebrews equated this star with Lilith, and in sixteenth-century Europe it was known by the Latin name *Caput Larvae*, "the specter's head." It was also called *Caput Medusae*, which means "the head of Medusa." Algol was one of Agrippa's important fifteen fixed stars.

Magical Interpretations and Uses for Perseus

As we have seen, Perseus has been equated with both sun gods Mithras and Lugh, making it the perfect constellation to incorporate into a Yule celebration. Lugh is the Shining One who represents the sun as nourisher of the earth. As for Mithras, his bull slaying represented a shift in the ancient night sky as the constellation marking the spring equinox changed from the bull (Taurus) to the ram (Aries). With the power to change the heavens, Mithras also represented the invincible sun. Whether Mithras, Lugh, or Perseus, this constellation represents the young god whose birth brings hope and light to the world.

Because we generally burn extra candles to mark our Yule sabbat, adding more to represent the Perseus star pattern may be less effective than desired. Instead, use a pine or fir bough trimmed to resemble the Y shape of the constellation. Alternatively, mistletoe berries can be used to lay out the star pattern. These serve a dual purpose as the white juice of the berries is often used to represent the seed of the god. After setting up your altar, draw down the energy of the Perseus constellation to enhance your sabbat ritual.

Some of my favorite Yule celebrations have taken place outdoors with the traditional jumping of a bonfire as we expressed our wishes for the coming year. Consider incorporating the following into your bonfire ritual. Trim small pine or fir boughs into the shape of the Perseus star pattern. Have enough for all participants to toss one into the fire when they make their jumps. The Perseus constellation is best seen during the month of December, so find him in the sky before jumping the bonfire and say: *"Perseus, Perseus, shining above; Increase light, hope, and love. As I jump this bright bonfire; Hear my wish and bring my desire."*

The Pleiades: Seven Sisters / Bringers of Peace and Wisdom
Messier Object: 45

Official Designation: 45M Tauri

Traditional Name: Pleiades

Pronunciation: PLEE-uh-deez

Visible Latitudes: 90° North to 65° South

Bordering Constellations: The Pleiades is an open cluster of stars within the Taurus constellation.

Description: Six of the seven stars form a semi-circle surrounded by a soft glow and nebulous haze.

To Find: Draw an imaginary line from the three bright stars of Orion's Belt northwest just beyond Aldebaran, the brightest star in Taurus. The next bright star is Alcyone in the Pleiades.

Figure 7.11. The Pleiades represent the powers of enchantment and wisdom.

The Pleiades is the brightest open cluster of stars in the sky. Unlike most star clusters, it has a discernible pattern and is one of the most identifiable celestial objects. Ancient depictions of these stars have been found in the Lascaux Cave in France and on the Nebra sky disc. An early reference to the Pleiades comes from China and dates to approximately 2357 BCE.

The earliest European written references are in a poem by Hesiod (circa 700 BCE) and in Homer's *The Iliad* and *The Odyssey*, written in the eighth century BCE.

Throughout time and in most cultures worldwide, the Pleiades have been observed, commented upon, and honored. Innumerable poets have written of their charm and mystery and described them as drops of dew, diamonds, pearls, and doves. The myth of the lost or invisible Pleiad is also universal, and it is believed that one of the stars may have been brighter in the past. The cluster contains hundreds of stars, but only six are easily visible to the unaided eye. You may see two other bright stars to the left of the Seven Sisters. These represent Atlas and Pleione, their parents.

In Greek mythology, the Pleiades were also known as the Atlantides, the daughters of Atlas, and were considered the virgin companions of Artemis. As the story goes, Orion saw the beautiful sisters and fancied them. After being pursued for seven years, Zeus answered the sisters' prayers for relief and transformed them into doves among the stars. These stars were also called Aphrodite's Doves. Both the astronomer Eudoxus and the poet Aratus considered the Pleiades a constellation and called it the Clusterers. The Romans called these stars the Bunch of Grapes. The Egyptians called them Atauria, which means "the stars of Athyr" (Hathor). The Latin translation of this name is Taurus.

Since very early times, the Pleiades served as calendar markers for plowing and harvesting. Abundant crops and green pastures were attributed to these Rainy Stars, as they were sometimes called. There is a wide range of legends from Native Americans, and many tribes referred to them as the Dancers. In addition, the Iroquois addressed prayers for happiness to the Pleiades. The Hindus of India called them the Flame of Agni, the god of fire. Additionally, in some Hindu myths they were equated with the seven wives of the Rishis or sages. In Japan they are called Subaru, which means "to unite." And yes, that starry emblem on these cars represents the Pleiades.

According to Ptolemy, the Pleiades are equated with Mars and the moon. Collectively, they were regarded as one of the important fifteen fixed stars.

The Stars of the Pleiades

Official Designation: Eta Tauri

Traditional Name: Alcyone

Pronunciation: al-SYE-oh-nee

This is the third-brightest star in the Taurus constellation and the brightest of the Pleiades. Alcyone is actually a five-star system. The primary component, Alcyone-1, is a blue-white binary star. Alcyone-2 and Alcyone-3 are white, and Alcyone-4 is yellow-white. At one time Alcyone was regarded as the central star in our galaxy, which the Arabs called the Bright One.

In one myth, Alcyone's liaison with Poseidon produced a daughter, Aethusa, who became the beloved of Apollo. In another myth, she and her husband Ceyx were transformed into kingfisher (halcyon) birds after they died. As a bird, Alcyone laid her eggs on the beach around the time of the winter solstice and her father, a god of the winds, subdued all breezes and calmed the waves to allow her peace. December 15 marks her feast day and the beginning of the halcyon days, which are the seven days before and after the winter solstice.

Official Designation: 16 Tauri

Traditional Name: Celaeno

Pronunciation: keh-LAY-no

This blue-white star is sometimes called the Lost Pleiad because it is the most difficult of the seven to find. Several other stars are also considered the lost sister. Despite Alcyone's liaison with Poseidon, in most legends it is Celaeno who married the sea god. Their union produced a son, Lycos, who ruled the Isle of the Blessed, which was also called Elysian Fields. In some stories it is said to be the lost continent of Atlantis. In other legends, Celaeno was associated with Prometheus.

Official Designation: 17 Tauri

Traditional Name: Electra

Pronunciation: ee-LECK-tra

This blue-white star is the third brightest in the cluster. In myth, her liaison with Zeus produced the son Dardanus, who became the founder of Troy. Electra married King Corythus of Tuscia, and with him bore a second son, Iasion. In some versions of Electra's story the paternity of her sons is switched. Nevertheless, when both sons died, her grief was so great that she wore a veil of mourning for the rest of her life.

Official Designation: 19 Tauri

Traditional Name: Taygeta

Pronunciation: TAY-geht-a

This is a triple star system in which Taygeta-1 and -2 are blue-white and Taygeta-3 is white. According to legend, Taygeta was unwilling to yield to Zeus and was disguised by Artemis as a female red deer to elude him. Of course, Zeus was a wily shape-shifter and took the guise of a stag. The outcome of their union was a son, Lacedaemon, who became the founder of Sparta. Taygeta was the city's patron goddess.

Official Designation: 20 Tauri

Traditional Name: Maia

Pronunciation: MY-ya

This blue-white star was named for the eldest of the sisters. First born and most beautiful, she was also the shyest. She was the third sister with whom Zeus dallied, and she gave birth to Hermes. Later, she became foster-mother to Arcas, son of Callisto, before they were placed in the heavens as Ursa Major and Ursa Minor. In Roman legend, she married Jupiter and was the mother of Mercury. She was also the goddess of their springtime festivals.

Official Designations: 21 Tauri and 22 Tauri

Traditional Names: Asterope; Sterope

Pronunciations: AS-ter-oh-pee; STER-oh-pee

This double star has the distinction of having Flamsteed numbers for each of its blue-white components. However, they share the same traditional names, which are derived from the Greek word for "star." Occasionally, a distinction is made with Asterope the name for 21 Tauri and Sterope 22 Tauri. In some legends, Ares was Asterope's lover and she gave birth to Oenomaus, who became king of Pisa. In other versions of her story, Oenomaus was her husband and they had a daughter named Hippodameia.

Official Designation: 23 Tauri

Traditional Name: Merope

Pronunciation: MURR-oh-pay

This blue-white star represents the only sister to marry a mortal, Sisyphus, who by some accounts was the founder of Corinth.

Magical Interpretations and Uses for the Pleiades

Through the ages, the Pleiades have been associated with mysticism and power. Some people claim that attunement to them aids in reaching a higher level of consciousness, while others use these stars to channel knowledge. For our purposes, we can use the Pleiades for a chakra alignment that will vitalize the flow of magical energy. It is also a nice way to start the New Year and set our intentions for the twelve months ahead.

We can activate and move chakra energy by using sounds. Conveniently, there are seven chakras within the body and seven Pleiades. Writer Amorah Quan Yin assigns the stars to the chakras as follows: Maia, root; Alcyone, sacrum; Electra, solar plexus; Celaeno, heart; Taygeta, throat; Asterope, third eye; and Merope, crown.[32] However, I suggest that three of them be switched for the following lineup: Merope, root; Maia, sacrum; Electra, solar plexus; Celaeno, heart; Taygeta, throat; Asterope, third eye; and Alcyone, crown. I put the powerful Alcyone at the crown chakra since this is our highest physical chakra and it leads to the celestial gateway. I put Merope at the root, as she was the sister who married an earthbound mortal. This left Maia in the position of the sacrum. As the goddess of Roman springtime festival, the sacral chakra, seat of creativity and procreation seemed the right place for her. That said, I also suggest trying it both ways or if your intuition leads you to a different arrangement, use it.

Lay out the Pleiades star pattern on your altar with tea light candles or dove-shaped glitter/confetti. Spend a minute or two to ground and center your energy. Chant the sound of Om three times. As you do this, focus your awareness on the energy of each chakra beginning at the root and moving upward from one to the next as energy flows supported from personal foundation into passion, into strength, and then into compassion and communication, to awareness and finally into cosmic consciousness. At the crown chakra, release your energy up through the three celestial chakras.

Next, draw the energy of the Pleiades down by chanting the star names. This time, start at the crown chakra and chant: "*Alcyone, Asterope, Taygeta, Celaeno, Electra, Maia, Merope.*" Do this slowly several times and visualize each chakra growing brighter as you chant the star names. Once you feel a shift in energy, cease chanting and sit in silence. Take time to simply be with the experience and open for any messages that the cosmos may send you. When it feels appropriate, end the session and ground your energy.

★ ★ ★

Taurus: The Bull/Horns of Consecration

Pronunciations: Taurus (TOR-us); Tauri (TOR-eye)

Visible Latitudes: 90° North to 65° South

Constellation Abbreviation: Tau

Bordering Constellations: Aries, Auriga, Cetus, Eridanus, Gemini, Orion, Perseus

Description: An elongated, sideways letter *V* with the open end pointing toward the east.

To Find: Draw an imaginary line from the three bright stars of Orion's Belt toward the northwest to Aldebaran, the brightest star in Taurus. The bull's horns are formed by the V shape.

This constellation's name is Latin for "bull." As noted in the entry for the Pleiades, it was derived from the name *Atauria*, which means "the stars of Athyr" (Hathor). Taurus has been known since at least the early Bronze Age, when it marked the sun's location during the spring equinox. Depictions of Taurus and the Pleiades have been found in Lascaux Grotto in France and date back to approximately 15,000 BCE. In Mesopotamia this constellation was known as the Bull of Light, and it was equated with the god Marduk. The bull was a prominent animal in cults of Babylon and Persia. Both Taurus and the Pleiades have been known in many indigenous cultures and referred to as the Bull and the Seven Sisters.

In Greek mythology, Taurus is associated with Zeus, who shape-shifted into a bull and offered himself as transportation for the Phoenician princess Europa. When she climbed onto his back, he abducted her and carried her to Crete. They eventually had three sons, one of whom became King Minos of Crete. Minos commissioned the palace at Knossos, where annually seven maidens and youths were sacrificed to the Minotaur, a creature that was half man and half bull. Archeological excavations of the palace revealed fabulous murals of bull-leaping games and horn emblems.

In Egypt, Taurus represented Apis, the bull of Memphis, who was a servant to Osiris. The constellation was also regarded as Horus, the son of Osiris, and represented the eternal return to life. Ptolemy equated the stars of Taurus with Venus, Saturn, Mercury, or Mars, depending on their location. Culpeper determined that this constellation influences the neck, ears, lower jaw, throat, and thyroid gland.

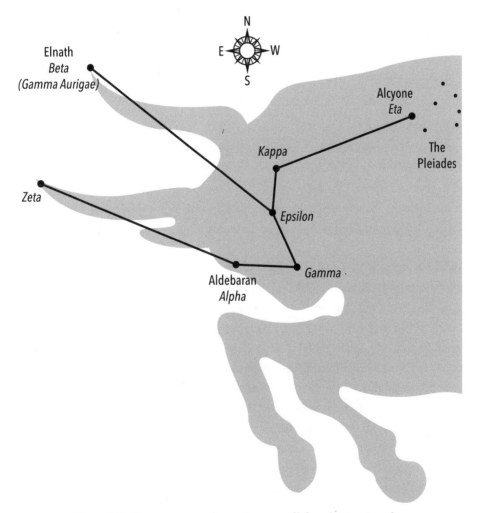

Figure 7.12. Taurus ensures the continuance of life and nature's cycles.

Notable Stars in Taurus

Official Designation: Alpha Tauri

Traditional Name: Aldebaran

Pronunciation: al-DEB-a-rahn

The traditional name of this orange star comes from Arabic and means "the follower" or "attendant." It was so called because of its proximity to the Pleiades. Also known as the Eye of the Bull, Aldebaran was one of the four royal stars of Persia and called the Guardian of the East. In addition, Aldebaran was one of Agrippa's important fifteen fixed stars. Ptolemy equated it with Mars.

Official Designation: Beta Tauri / Gamma Aurigae

Traditional Name: Elnath

Pronunciation: EL-noth

This blue star is common to Taurus and Auriga the Charioteer, where it is the gamma star. In Taurus it is positioned at the tip of one of the bull's horns. Its traditional name comes from Arabic and means "the butting one." It was also called the Bull's North Horn. Along with the Pleiades, the Hindus associated this star with Agni, the god of fire. According to Ptolemy, Elnath is equated with Mars and Mercury.

Magical Interpretations and Uses for Taurus

As the paintings in the hall of bulls of Lascaux Grotto attest, this animal has been vital to humans for food as well as an important symbol. The bull represented power, strength, and potency for many thousands of years. It personified the forces of the sun, representing fertility and abundance. The bull was sacred to the Great Goddess, and later it became a symbol of power for the mightiest gods such as Amun, Jupiter, Mars, Mithras, Osiris, Poseidon, Shiva, and Zeus. The bull was an emblem of life, and its blood represented regeneration. Bull slaying was believed to release the life-giving forces of the animal, ensuring the continuance of nature's cycles. The symbolism of power and protection rested within the bull's horns, which became the sacred horns of consecration. On the palace of Knossos these were represented by a stylized U shape, which adorned the top of the palace and other places throughout the complex.

In a similar vein, altars in the temples of Luxor and Karnak in Egypt had raised corners, which were called the horns of the altar. The altars of the Hebrews also had horns, and these were considered the most sacred part. Using the energy of the Taurus constellation, we can consecrate and empower our altars as sacred space. Also, just as offerings took the place of sacrifice, we can use the star pattern of Taurus to honor deities and the eternal cycles of nature.

Clear everything from your altar and then burn frankincense or myrrh to begin the process of consecrating your space. Waft the smoke over and around the altar, and then set the incense aside. Light one red candle and set it on the opposite side of the altar from where you are standing or sitting. Lay out the Taurus star pattern in the center of the altar using pieces of lapis lazuli, and/or malachite, and three pieces of red carnelian. Draw down the star energy and then hold your hands above the gemstones as you say: *"Mighty Taurus, bull of the heavens, I call on you to empower this altar and make it worthy to serve the Great Mother Goddess and the gods of power. Make this sacred, consecrated space. As above, so below."*

After releasing the energy into the gemstones, take the star pattern apart and move one piece of lapis or malachite to each corner to symbolically become the horns of the altar. Move the candle to the center of the altar, and then place the three pieces of carnelian in front of it to symbolize the blood of the bull as you say: *"Life-giving energy of Taurus, of the Great Mother Goddess and the God. May the Wheel of the Year turn and cycles continue in the never-ending spiral of life. So mote it be."* Take some time to honor your special deities, and then leave the gemstones in place for three days. Put them back on your altar during ritual or magic work for energetic support.

Chapter Eight
THE SOUTHERN HEMISPHERE

A number of southern constellations cataloged by Ptolemy were already covered in the preceding seasonal chapters because of their visibility in the Northern Hemisphere. Those that are farther south were "unseen" or not readily noticed by Europeans. As a result, these constellations were not mapped until the fifteenth century. However, most of the constellations introduced in this chapter can be seen in parts of the Northern Hemisphere.

Most of the far southern constellations were created during the age of exploration when many areas of the earth and skies were being mapped for the first time. Dutch astronomer and cartographer Petrus Plancius instructed navigators Pieter Dirkszoon Keyser (1540–1596) and Frederik de Houtman (1571–1627) on how to observe the night sky. The stars that they documented were divided into twelve constellations, which Plancius published on a globe in 1598. Plancius was gratified when the famous Johann Bayer included these constellations in his 1603 star atlas. In this chapter, we will look at a few of Keyser and Houtman's constellations, many of which depict animals.

French astronomer Nicolas Louis de Lacaille (1713–1763) observed and catalogued more than nine thousand stars in the Southern Hemisphere during his two-year stay in Cape Town, South Africa. From this work he created thirteen new constellations. As today, technology was a hot topic and in keeping with the scientific enthusiasm of his day,

Lacaille named most of his constellations for instruments and devices. In some cases, he "moved" stars from known constellations into his new ones.

Because a number of southern constellations have already been covered, this chapter makes note of seasonal differences in their use as well as practices that require slight alteration. Reference is made for the chapters in which the particulars, history, and magical uses of these constellations can be found. Because locating these constellations from the Southern Hemisphere is different from in the north, details on how to find them have been adjusted. Also, as in the previous chapters, directions given to locate constellations and stars assume that the reader is facing south.

While the constellations introduced in this chapter are not directly associated with classical myths and do not have lengthy historical backgrounds, they do have relevance for ritual and magic. In the southern skies we will find two birds (a crane and a phoenix), a fish, and a cross. As in the north we also find a snake and a crown. Let's take a look at the southern sky season by season.

The Southern Spring Sky: September, October, November

Refer to chapter 6 for the particulars and history of spring constellations that were covered in the Northern Hemisphere.

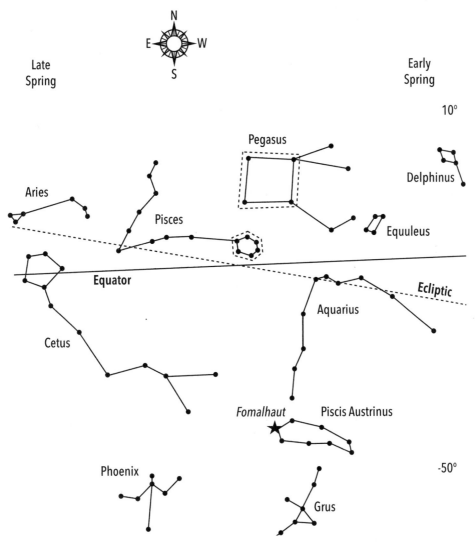

Figure 8.1. The spring night sky showing constellations and the brightest star, Fomalhaut. The Great Square of Pegasus and the Circlet of Pisces asterisms are indicated with dotted shapes.

Aquarius: The Water Bearer/Sea God

To Find: Look for the bright star Fomalhaut in Piscis Austrinus. Aquarius is just to the north.

While a number of northern sea gods were mentioned in chapter 6, I leave it to readers in the Southern Hemisphere to choose gods of Oceania that may better suit their locations and traditions. Also, as you prepare your garden in the spring, call on Aquarius to send nourishing water. Draw the star pattern on a fence or on a couple of stones that you can secret away under some plants. If you have a fountain or water feature, place an image or pendant of Aquarius on it or somewhere nearby to draw this constellation's influence. Because a garden may not be the only thing you want to nurture, call on Aquarius for support in developing talents and skills. In addition, with a number of stars in this constellation associated with luck, the energy of Aquarius can give good luck spells a boost.

Aries: The Ram/Power of the Horned God

To Find: Look for the bright star Fomalhaut in Piscis Austrinus. Draw an imaginary line toward the northeast and follow the stars of Cetus. Aries is just north of the circle of stars that form the whale's head.

Call on the energy of Aries to connect with the Horned God, to enhance dark moon rituals, and to boost the power of spells especially those for protection.

Cetus: The Whale/Keeper of Traditions

To Find: Look for the bright star Fomalhaut. Cetus is to the northeast of Piscis Austrinus and east of Aquarius. It stretches underneath Pisces.

This whale brings balance to equinox sabbat rituals. In addition, Cetus reminds us that we can always make changes in our lives. This constellation can be instrumental in moving energy toward the changes we seek.

Delphinus: The Dolphin/Carrier of Souls

To Find: Look for the bright star Fomalhaut in Piscis Austrinus and draw an imaginary line to the northwest through Aquarius and the trapezoid of Equuleus. Angle that line a little to the west and north to the kite shape of Delphinus.

In a number of cultures, the dolphin is associated with carrying souls to the land of the dead. Although this part of the entry in chapter 6 is written for Samhain, it is applicable anytime for support when a loved one passes to the other side of the veil.

Grus: The Crane/Seeker of Knowledge

Pronunciations: Grus (groos); Gruis (GROO-iss)

Visible Latitudes: 34° North to 90° South

Constellation Abbreviation: Gru

Bordering Constellations: Phoenix, Piscis Austrinus

Description: A line of stars curves gently from north to south. Three of its brightest stars form a triangle.

To Find: Locate the bright star Fomalhaut in Piscis Austrinus. Grus is just to the south.

Originally named Krane Grus, which means "crane" in both Dutch and Latin, only the Latin was used when it became an official constellation in 1922. Also, *Grus* is the genus name for cranes. Like many of the constellations in the Southern Hemisphere, Grus was mapped by Dutch navigators and made public by astronomer Petrus Plancius in the late 1590s. For a brief time during the seventeenth century in England, this constellation was called Phoenicopterus, which means "flamingo" in Latin.

Ptolemy originally mapped these stars as part of Piscis Austrinus and equated them with Venus and Mercury. There are no legends associated with this constellation; however, Celtic myth abounds with stories and references to cranes.

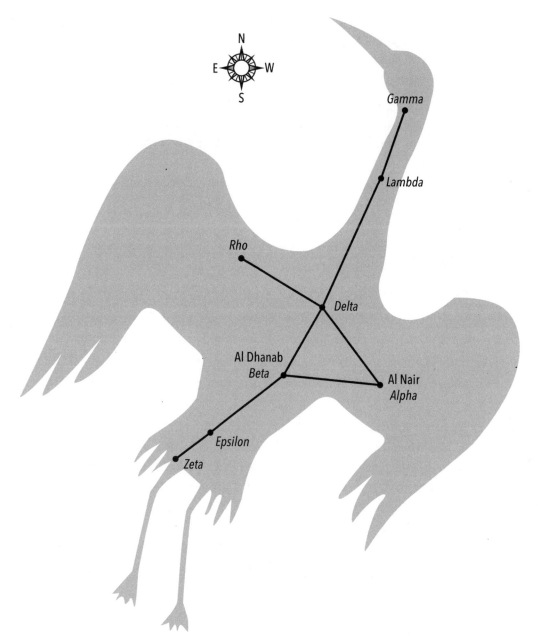

Figure 8.2. Grus is the purveyor of ancient wisdom.

Notable Stars in Grus

Official Designation: Alpha Gruis

Traditional Name: Al Nair

Pronunciation: al NAYR

The traditional name of this star is Arabic and means "the bright one." It was originally "the bright one in the fish's tail" and a holdover from when this star was considered part of Piscis Austrinus. On Grus, this star is located on the bird's wing. Al Nair is a blue-white star.

Official Designation: Beta Gruis

Traditional Name: Al Dhanab

Pronunciation: al DAH-nab

The traditional name of this red star comes from Arabic and means "the tail," which also dates to its time as part of Piscis Austrinus. On Grus, this star is located on the bird's body. The alternate spelling "Deneb" can be found in other star names in the Capricornus, Cetus, Cygnus, and Delphinus constellations.

Official Designation: Delta Gruis

This is a set of double stars. Delta-1a is yellow, and its companion Delta-1b is orange. The Delta-2 double stars are both red. These stars mark the bird's breast.

Magical Interpretations and Uses for Grus

Found throughout Celtic myth, the crane is associated with death, rebirth, understanding deep mysteries, and truth. Many legends tell of women shape-shifting into cranes. The famous crane bag of the sea god Manannán mac Lir was made from such a bird when the woman died. The bag that Manannán created and used was said to hold magical objects and things of power. The forfeda, the fifth group of characters in the Ogham alphabet, is sometimes called the Crane Bag and was said to have been created by the sea god.

Grus is the perfect constellation for energetically charging a bag or other container that holds your divination tools. Additionally, buy or make a drawstring bag for your divination tools and then embroider or draw the Grus star pattern with fabric marker. When it is finished, place it on your altar and draw down the energy of Grus, saying: *"Powerful crane, starry Grus; Bless this bag for special use. Empower the tools that it will keep; And hold the knowledge that I seek."*

To give your divination or shamanic practices a boost, lay out the Grus star pattern using images of the Ogham forfeda characters. Hold your divination tools or items that you use for shamanic work as you stand in front of your altar. Draw down the energy of Grus as you focus attention on your third eye chakra. When the energy increases to a crescendo, release it into your tools and send any excess energy down through your earth star chakra.

In Celtic myth, a ritual crane posture of standing on one leg with one eye closed was said to have been employed by Lugh and the Dagda. While this was used for general spellwork, it was also instrumental in performing a type of magic known as *corrguinecht,* "crane-wounding." I mention this not to advocate the use of retributive magic, but to illustrate the power associated with this bird. For our purposes in star magic, the crane posture can be used while drawing down the energy of the Grus constellation before or during an Ostara ritual or magic work.

Pegasus and Equuleus: The Winged Horse and the Colt/Powers of Nature
To Find: From the star Fomalhaut in Piscis Austrinus, draw an imaginary line north to the
   Great Square of Pegasus. Equuleus is located in front (to the west) of Pegasus.

Just as Pegasus is used in the north at the autumn equinox, so too is it appropriate for the spring. The winged horse is a symbol of the heightened power of the natural forces, and the Great Square of Pegasus emphasizes this by representing the four cardinal directions. It also represents the four points of the solar year and the elements. Pegasus is an aid for divination and other psychic practices as well as a vehicle and guide for astral travel.

Phoenix: The Phoenix/Energy of Self-Transformation
Pronunciations: Phoenix (FEE-nicks); Phoenicis (feh-NYE-siss)
Visible Latitudes: 32° North to 90° South

Constellation Abbreviation: Phe

Bordering Constellations: Eridanus, Grus

Description: A horizontal zigzag pattern forms the bird's wings and is intersected by a vertical line that forms the body.

To Find: Look for the bright star Fomalhaut in Piscis Austrinus, and the Phoenix is to the southeast. It is east of Grus.

Like many southern constellations, the Phoenix was mapped by Dutch navigators and published by astronomer Petrus Plancius. Although there is no classical myth per se associated with it, this constellation depicts a mythical bird that was known in many ancient cultures. The bird was said to resemble an eagle, but with red, gold, and purple feathers making it far more flamboyant and impressive.

According to Greek, Roman, and Arab legends, the bird had a life span of five hundred years. When it reached the end of its days, it built up its nest into a funeral pyre. Once the nest was ignited by the sun's rays the phoenix went up in flames, but then it rose from the ashes and began a new life cycle. In his work *The Metamorphoses*, Roman poet Ovid (43 BCE–17 CE) explained that the phoenix built a nest of bark, cinnamon, and myrrh atop a palm tree. As the burning incense lifted the soul of the old phoenix away, a new little bird emerged from the old one's body.

In Chinese legend, the phoenix lived from one hundred to one thousand years. Known as the magical Firebird, it was a symbol of good luck. Although the Arabs knew the bird in legend, they did not connect it with this constellation and instead called these stars the Boat.

Notable Stars in Phoenix

Official Designation: Alpha Phoenicis

Traditional Name: Ankaa

Pronunciation: ANG-kah

The traditional name of this star comes from Arabic and means "the phoenix." This orange, spectroscopic binary star marks the neck of the phoenix. Ankaa has been linked with ambition and potential fame.

Official Designation: Beta Phoenicis

This yellow binary star does not have a traditional name. It marks one of the bird's wings.

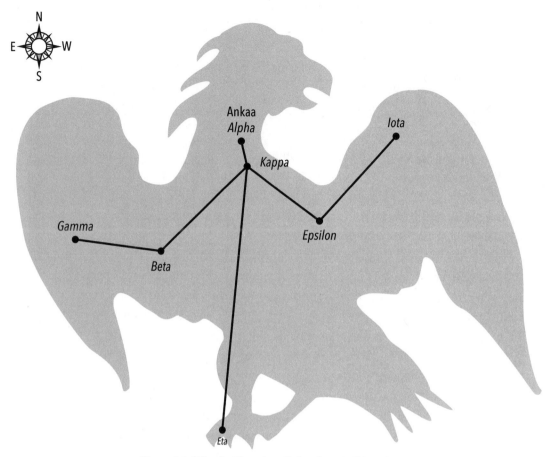

Figure 8.3. Like the Phoenix, all that dies shall be reborn.

Magical Interpretations and Uses for the Phoenix

Quite obviously, this constellation is about change and cycles. It is appropriate for an Ostara altar to represent the change in season as well as to symbolize Persephone's reemergence from the underworld. Lay out the star pattern on your altar with tea light candles to emphasize the fire aspect of the Phoenix. Anoint the candles first with a little cinnamon oil or burn myrrh while you set up your altar. Draw down the energy of the constellation and release it into your altar to enhance your sabbat ritual.

Spring is a time of metamorphoses as the world renews its beauty and splendor. Nature invites us to not only open our windows, but also open ourselves for personal growth. Because in legend the transformation of the phoenix is a solitary act in which the bird gives birth to itself, this constellation can provide support when we need to initiate change and move our lives forward. This exercise will be done during the day, but prepare for it the night before. You will need seven gemstones; a mix of amethyst, red garnet, and citrine. Lay out the Phoenix star pattern outside on a porch or inside on a windowsill. When it is dark and the stars are shining, draw down the energy of the constellation and release it into the stones.

The next day, go outside where you can sit without being disturbed. Take the gemstones and a beach towel with you. Lay out the star pattern on the towel in front of you, close your eyes, and hold your palms over the gemstones as you say: *"Stars of beauty, bird of change; Help my life to rearrange. Fire burn as I pass through; Like a phoenix I will renew."* Visualize the energy of the constellation moving from the stones, up your arms, and throughout your body. Imagine yourself as a magnificent phoenix and visualize all the things that you want to let go of burning away around you. Feel the heat consume you and the energy lift you. Feel your fiery wings expand as you rise, knowing that whatever challenges are ahead you will meet them with strength and poise.

Allow the image to fade as you slowly bring your attention back into your physical body. Bring the energy down to your earth star chakra until you feel grounded and stable. Spend a few minutes gazing at the world around you with new eyes. You have been through the fire of change and you will now blossom and grow.

Pisces: The Fish/Duality, Unity, and Divination

To Find: From the star Fomalhaut in Piscis Austrinus, draw an imaginary line north and a
little east between Aquarius and Cetus.

The *Vesica Pisces*, "vessel of the fish" associates Pisces with procreation and creativity. These are especially apropos for spring as the world blossoms with new life. A knotted cord joins the two fish depicted in this constellation. A knot, of course, is the symbol of a bond. Although the entry in chapter 6 for this association was written with Samhain in mind, like Delphinus the energy of this constellation can be employed whenever a loved one passes to the other side of the veil.

Piscis Austrinus: The Southern Fish/Ancient Wisdom

Pronunciations: Piscis Austrinus (PIE-siss awe-STRY-nus);

    Piscis Austrini (PIE-siss awe-STRY-nye)

Visible Latitudes: 55° North to 90° South

Constellation Abbreviation: PsA

Bordering Constellations: Aquarius, Capricornus, Grus

Description: A crooked rectangular shape.

To Find: Look for Fomalhaut, the brightest star in the spring sky.

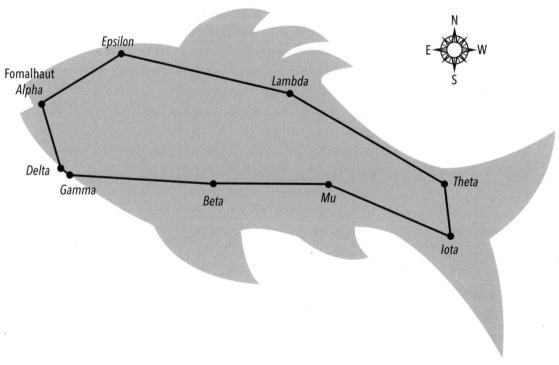

Figure 8.4. Piscis Austrinus is a symbol of otherworldly knowledge.

Also called Piscis Australis, this constellation's name means "the southern fish" in Latin. The Babylonians simply called it the Fish. In Egypt, this constellation was also regarded as a fish and associated with a legend about Isis being rescued by a fish. The Greeks called it the Great Fish and depicted it floating on its back swallowing the water poured from the jar of Aquarius. The two fish of the Pisces constellation were said to be its offspring.

In the late 1590s when Dutch astronomer Petrus Plancius was mapping the new constellations created by explorers, he took some stars away from Piscis Austrinus to form Grus. Additionally, for a period of time in the 1800s, it was known as Piscis Notius, which means "famous fish."

Notable Stars in Piscis Austrinus

Official Designation: Alpha Piscis Austrini

Traditional Name: Fomalhaut

Pronunciation: FOE-mal-haht

The name of this white star comes from an Arabic phrase that means "the mouth of the fish." It was one of the four Persian royal stars and known as the Guardian of the South. Ptolemy equated it with Venus and Mercury. Throughout time, Fomalhaut has been considered a fortunate and powerful star. Fomalhaut is an amplifier of energy.

Official Designation: Epsilon Piscis Austrini

This blue-white star is the second brightest in the constellation. It marks the top of the fish's head.

Magical Interpretations and Uses for Piscis Austrinus

A fish and its watery realm are symbolic of the subconscious, the mysterious unknown, and knowledge that comes from deep sources. In Celtic legend, a fish in the form of a salmon was considered one of the oldest and wisest of creatures. It served as a symbol of wisdom and otherworldly knowledge, and it was associated with prophecy, divination, and inspiration. The salmon was said to have received knowledge by eating the hazelnuts of the nine trees of wisdom that fell into the water. When you engage in divination, psychic work, or shamanic travel, lay out the Piscis Austrinus star pattern with hazelnuts and draw down the energy of this constellation to boost your practices. Since Fomalhaut is a bright and powerful star, use a piece of white quartz in place of a hazelnut to mark its position in the star pattern.

THE SOUTHERN SUMMER SKY: DECEMBER, JANUARY, FEBRUARY

Refer to chapter 7 for the particulars and history of summer constellations that were covered in the Northern Hemisphere.

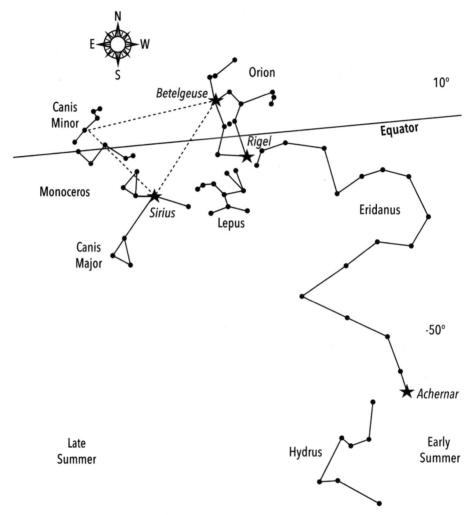

Figure 8.5. The southern summer sky showing four of the brightest stars.
The Winter Triangle asterism of the Northern Hemisphere
becomes the Summer Triangle in the Southern Hemisphere.

Canis Major and Canis Minor: The Great Dog and the Little Dog/
Guardians and Guides

To Find: Look for Sirius, the brightest star in the sky, located in Canis Major. The dog's body stretches to the southeast, and his front leg is to the west of Sirius. To find Canis Minor, draw an imaginary line from Sirius to the northeast to the next bright star, Procyon, which is the little dog's alpha star.

Throughout time and around the world, dogs have served as guardians. The Babylonians considered even the image of a dog to be magically protective. Refer to the main entry on these constellations for details on calling on the celestial dogs as guardians. In some of the very ancient cultures of Europe, dogs represented the energy of spring. They were the guardians of life who oversaw the growth of vegetation. Fulfilling their dual nature, Canis Major and Minor take on the qualities of Canes Venatici, the northern dogs of spring. Refer to chapter 4 for information on their influence for gardens.

Eridanus: The River/Flowing Sacred Waters

To Find: Locate Orion's Belt and then look southwest to the bright star Rigel that marks a foot of the hunter. Cursa, the beta star in Eridanus, is to the west of Rigel. Follow the trail of stars west and then south. The constellation meanders east and then south again.

In many cultures, rivers were thought to have special powers or were used for worship. As a result, a number of goddesses became associated with rivers. These goddesses and their waterways represent power, wisdom, and transformation. We can call on the energy of river goddesses or any river deity with the help of the celestial river Eridanus for ritual, spellwork, and meditation.

Hydrus: The Southern Water Snake/Ancient Goddess Energy
Pronunciations: Hydrus (HIGH -drus); Hydri (HIGH-dry)

Visible Latitudes: 8° North to 90° South

Constellation Abbreviation: Hyi

Bordering Constellation: Eridanus

Description: A triangle asterism of its alpha, beta, and gamma stars is easiest to see. The other stars in the constellation form a zigzag pattern through the triangle.

To Find: The head (alpha star) of Hydrus is located just below and to the southeast of Achernar, the last star in Eridanus.

Hydrus is one of two circumpolar constellations in the southern sky covered in this book. The name of this constellation is derived from Greek and means "the water snake." It is sometimes called the Lesser Water Snake in regard to the larger Hydra constellation in the north. There is no mythology directly associated with this constellation. It was mapped by Dutch navigators and was first published by astronomer Petrus Plancius. Although French astronomer Nicolas Louis de Lacaille thought it should be known as *l'Hydre Mâle*, "the male Hydra," gender distinction is not often made.

Notable Stars in Hydrus

Official Designation: Alpha Hydri

Official Designation: Beta Hydri

Official Designation: Gamma Hydri

The stars in Hydrus do not have traditional names. Alpha Hydri is a yellow-white star and the second brightest in the constellation. It marks the head of the snake. The yellow Beta Hydri is the brightest in the constellation. Gamma Hydri is a red star and forms a triangle with the other two.

Magical Interpretations and Uses for Hydrus

The snake is one of the oldest symbols of the Great Mother Goddess. She represents the principle of life, reproduction, and the forces of the natural world. All of the attributes and powers of Hydra noted in chapter 4 can be applied to Hydrus.

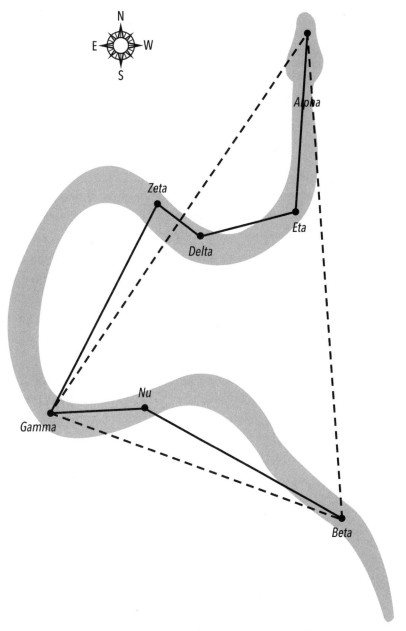

Figure 8.6. Hydrus is a symbol of the Great Goddess and her power.

Lepus: The Hare/Magical Moon Energy

To Find: Lepus is directly south of Orion and west of the bright star Sirius in Canis Major.

We can call on the energy of Lepus for aid in divination practices. In addition, because the hare is associated with the moon, the energy of this constellation can enhance esbat rituals and boost moon magic. It also aids in working with the energy of the new moon.

Monoceros: The Unicorn/Power of the Crone

To Find: Look for Sirius, the brightest star in the sky, located in Canis Major. Draw an imaginary line from Sirius to the northeast to the next bright star Procyon in Canis Minor. Monoceros is between these two constellations.

Monoceros can be called upon to deepen spiritual meditations, to stoke the flame of creativity, and to initiate self-transformation. This symbol of beauty, enchantment, love, and wisdom carries the power of the crone.

Orion: The Hunter/Artemis the Huntress

To Find: Located northwest of Sirius, the distinctive arrangement of three bright stars in a straight line marking Orion's Belt makes this constellation easy to find. The bright star Rigel, southwest of the belt, represents the hunter's left foot.

Quite naturally, this constellation represents male hunter gods. In addition, legends of the spectral Wild Hunt can be equated with furious storms. These storms can be used to boost a spell, or they can be an opportunity to experience the power of the natural world. Because her story is intertwined with Orion's, I like to think of this constellation as Artemis herself. She can help us find our strengths and independence to live as we choose. In addition, as an archer she can empower us to target and reach our goals.

THE SOUTHERN AUTUMN SKY: MARCH, APRIL, MAY

Refer to chapter 4 for the particulars and history of autumn constellations that were covered in the Northern Hemisphere.

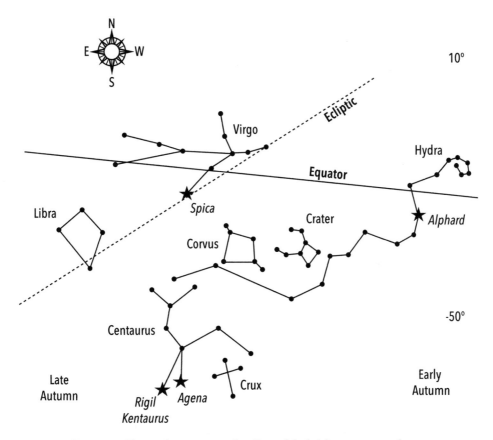

Figure 8.7. The southern autumn sky. Four of the brightest stars are shown.

Centaurus: The Centaur/Chiron the Healer

To Find: Look for the two bright alpha and beta stars, Rigil Kentaurus and Agena, respectively. These mark the front feet of the centaur and are located east of Crux the Southern Cross.

The bright alpha and beta stars of this constellation are called the Pointers in the Southern Hemisphere because they point toward Crux. Centaurus presents us with duality and balance. When we find ourselves struggling with two aspects of self, we can call on this constellation for guidance. Because Centaurus represents Chiron, we can also call on these stars to boost healing energy.

Corvus: The Crow/Messenger from Other Realms

To Find: Locate the two bright stars that mark the front feet of Centaurus and draw an imaginary line north and slightly west through Hydrus. The next constellation is Corvus.

The crow is considered to be on the edge between light and dark, life and death. Because crows are scavengers and feed on dead things, they were considered messengers from the otherworld. This association is particularly apropos for divination at Samhain. Like many of the spring and autumn constellations, Corvus has a dual nature. To the Greeks, crows functioned as messengers of the gods bearing wisdom and secrets. We can call on Corvus to aid us in divination as well as otherworld journeys.

Crater: The Cup/Chalice of the Goddess

To Find: Locate the two bright stars that mark the front feet of Centaurus and draw an imaginary line northwest through Hydrus. Crater is just west of Corvus, tucked into a curve of Hydra the Water Snake.

The cup, or chalice, is the magical and ritual tool for the element water. The cup represents the vessel of plenty, and on our altars it represents the Goddess. We can use it to receive the energy that we draw down from this constellation to amplify her presence. The energy of Crater can also be used to bless your altar and ritual space. Because the cup/chalice represents a womb, Crater is an appropriate constellation for spellwork relating to fertility.

Crux: The Southern Cross / The Power of Four

Pronunciations: Crux (krucks); Crucis (CREW-siss)

Visible Latitudes: 20° North to 90° South

Constellation Abbreviation: Cru

Bordering Constellation: Centaurus

Description: Four bright stars form an easily recognizable cross-shaped pattern.

To Find: Although Crux is bright and well known, it can be confused with the False Cross
asterism, which is not as bright, but larger. The way to check that you have found the
correct cross is to look for the Pointers, Rigil Kentaurus and Agena in Centaurus. Crux
is located under the body of the Centaur.

Although Crux is the smallest of the eighty-eight constellations, it is one of the best known
in the Southern Hemisphere. Its name means "the cross" in Latin. Crux is one of the two
southern circumpolar constellations covered in this book. While there are no bright stars
that mark the south celestial pole, Crux points to the pole's location.

The earliest recorded notation about Crux was made by the Greeks, who regarded it
as part of Centaurus. However, it was so low in the sky that by 400 CE the precession of
the equinoxes had pushed it out of sight below the horizon. Those in the north forgot
Crux until Europeans began exploring the southern regions in the fifteenth century. In
1679, French astronomer Augustin Royer gave Crux its due and made it a constellation
independent from Centaurus.

To Australian Aboriginal people, Crux forms part of the head of a figure they call the
Emu. In addition, Crux is featured on the national flag of Australia. The Inca of Peru called
this constellation Chakana, which is the name of their sacred stepped cross.

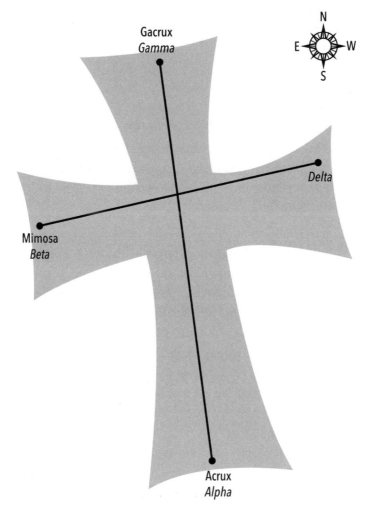

Figure 8.8. Crux represents the power of the four directions and elements.

Notable Stars in Crux

Official Designation: Alpha Crucis

Traditional Name: Acrux

Pronunciation: AY-krucks

Acrux is a multiple star system composed of a spectroscopic binary (Alpha-1) and a single star (Alpha-2). All three are blue-white. Acrux is the southernmost star in the constellation. The name Acrux is an abbreviation of its designation as the alpha star and the name of the constellation (A'Crux). Ptolemy equated this star with Jupiter. It has been associated with magic and mystery.

Official Designation: Beta Crucis

Traditional Names: Mimosa; Becrux

Pronunciations: mim-OH-sah; BAY-krucks

Mimosa is a blue-white, spectroscopic binary star that marks the eastern (left) arm of the cross. The reason for this name is unknown. Some sources indicate that it was derived from the Latin *mimus*, which means "actor" or "mime." Others say that it is related to the star's color, which is also a vague interpretation since mimosa blossoms are usually pink or lavender and white. This star is sometimes referred to as Becrux, an abbreviation of its beta designation and the name of the constellation.

Official Designation: Gamma Crucis

Traditional Name: Gacrux

Pronunciation: GAH-krucks

Like the other two stars, the name of this one is an abbreviation of its Greek letter and the name of the constellation. Unlike the others in this constellation, Gacrux is a red star. It marks the northernmost star in the cross.

Official Designation: Delta Crucis

This blue-white star marks the western arm of the cross. Unlike the other three stars that form the cross, this one has not been named. It is a variable star that has a slight change in brightness approximately every four hours.

Magical Interpretations and Uses for Crux

Pagans and Wiccans should not shy away from this constellation because it is a symbol of Christianity. In other times, crossed lines represented the four directions and the four phases of the moon. It also served as a symbol of life and elemental energy. Use these associations to bring power to your rituals. Position the Crux star pattern on your altar so that your direction/element candles are in the spaces between the arms of the cross.

Either before your ritual or prior to lighting the candles and calling in the directions, draw down the energy of Crux. Release it into your altar as you follow the pattern of the cross with your finger and say: *"Crucis, Crucis, your pattern I trace; Bring your energy to this space; As I call the directions and deity; Shine brightly above, so mote it be."*

Because Crux also represents a crossroads, we can use the power of this constellation for support when we are at a turning point in our lives. Take a piece of paper and cut out a small cross with the arms wide enough to write a few words upon. On a starry night, spend a few minutes thinking about whatever crossroad or change you are facing, and then write a couple of keywords about it on the paper cross. Go outside, locate Crux, and hold the paper between your hands. Draw the star energy into the paper as you say: *"From above, stars of Crux; Help me through this time of flux. Shine your light so I may see; The path ahead that's best for me."* When you go back inside your house, place the paper cross under a candle on your altar. Leave it in place until change has occurred or your crossroad has been resolved.

Hydra: The Water Snake/Symbol of the Goddess and Transformation
To Find: Locate the two bright stars that mark the front feet of Centaurus and draw an
   imaginary line north. The tail of Hydra is above Centaurus. The snake winds to the
   west and north. Look for the bright star Alphard and a compact group of stars to the
   northwest of Alphard that marks the head of Hydra.

It seems appropriate to have one of the most ancient symbols of the Great Mother Goddess, the snake, balancing the night sky in the spring (Hydrus) and the autumn (Hydra). Call on this constellation to represent the power of the Goddess for Mabon and Samhain rituals.

Libra: The Scales/Balance and Justice
To Find: Locate the two bright stars that mark the front feet of Centaurus and draw an
   imaginary line to the northeast. Libra is the next constellation in that direction.

Representing weighing scales, Libra is a reminder that balance is important in our lives. The energy of this constellation fits well with autumn equinox rituals. We can call on Libra for aid in bringing our health and the energy of our homes into balance, too. Additionally, just as ancient people equated Libra with the goddesses of justice, we can call on the power of this constellation for support in legal matters or whenever we need to right a wrong.

Virgo: The Virgin/Maiden and Mother Nurture the World

To Find: Locate the two bright stars that mark the front feet of Centaurus and draw an imaginary line north through Hydrus and Corvus. Virgo stretches above Corvus and Crater. Look for Spica, just above Corvus. It is the brightest star in Virgo and one of the brightest in the night sky.

In the Northern Hemisphere, Virgo represents the goddess of fertile soil and spring's renewal. In the Southern Hemisphere, she is the lady of the harvest bringing abundance to our Mabon rituals. If you use wine in your celebrations, call on the Star of Bacchus, Vindemiatrix, to pour forth special blessings.

THE SOUTHERN WINTER SKY: JUNE, JULY, AUGUST

Refer to chapter 5 for the particulars and history of winter constellations that were covered in the Northern Hemisphere.

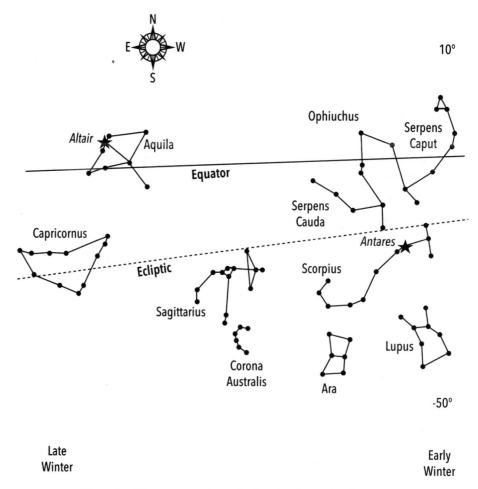

Figure 8.9. The southern winter sky showing the two brightest stars.

Aquila: The Eagle/Power of the Sun

To Find: Locate the bright star Antares in Scorpius and draw an imaginary line east and slightly north. Look for the bright star Altair, which is the middle of three stars across the eagle's back.

While the eagle embodies the spirit of summer in the north, here in the Southern Hemisphere it represents the return of the sun at winter solstice. Call on the power of Aquila for Yule as well as for honoring the powerful gods who are associated with the eagle.

Ara: The Altar/Heavenly Sacred Space

To Find: Locate the bright star Antares in Scorpius. A line of stars that trail to the southeast delineate the scorpion's tail. Ara is below the tail.

Ara serves as a reminder of the importance of an altar as a place to give thanks both indoors and out. An outdoor altar is a good place to leave food for birds or other animals to help them get through the winter. With the Latin meaning of *ara* being "a place of refuge and protection," the energy of this constellation can be used to charge a protective talisman for your house or land.

Capricornus: The Sea Goat/Horned One of Abundance

To Find: Locate the bright star Antares in Scorpius. Capricornus is east on the other side of Sagittarius.

Linked with Pan and, quite naturally, the Horned God, call on Capricornus to celebrate him at Yule. In addition, because Capricornus is associated with the horn of plenty, call on him for abundance through the winter months or to boost prosperity spells.

Corona Australis: The Southern Crown / Heavenly Heart

Pronunciations: Corona Australis (cuh-ROE-nuh awe-STRAL-iss);

Coronae Australis (cuh-ROE-nee awe-STRAL-iss)

Visible Latitudes: 40° North to 90° South

Constellation Abbreviation: CrA

Bordering Constellations: Ara, Sagittarius, Scorpius

Description: The stars form a graceful semicircle.

To Find: Locate the bright star Antares in Scorpius. A line of stars that trail to the southeast delineate the scorpion's tail. Continue that line to the southeast between Sagittarius and Ara to Corona Australis.

The name Corona Australis is Latin for "the southern crown," as opposed to the northern constellation of Corona Borealis, which means "the northern crown." This constellation is sometimes called Corona Austrina. Located near the front feet of Sagittarius, it was formerly known as Corona Sagittarii, "the crown of Sagittarius." Instead of a jeweled crown, the ancient Greeks regarded this constellation as the type of laurel wreath used to bestow honors on athletes and other people of talent. In addition, Corona Australis is sometimes associated with Semele, the mother of Dionysus, and commemorates his rescue of her from the underworld.

Ptolemy equated the energy of this constellation with Saturn and Jupiter.

Notable Stars in Corona Australis

Official Designation: Alpha Coronae Australis

Traditional Name: Alphecca Meridiana

Pronunciation: al-FECK-ah mer-ID-ee-ah-nah

This white star is located in the upper left of the semicircle and in between two of the other brightest stars (beta and gamma). Its traditional name is a mix of languages. *Alphecca*, which is also the name of the alpha star in Corona Borealis, comes from Arabic and means "break" or "broken," referring to the open or broken circle of stars that form the constellation. *Meridiana* is Latin and means both "south" and "midday." In this case, it means "south" to distinguish this star from its northern counterpart.

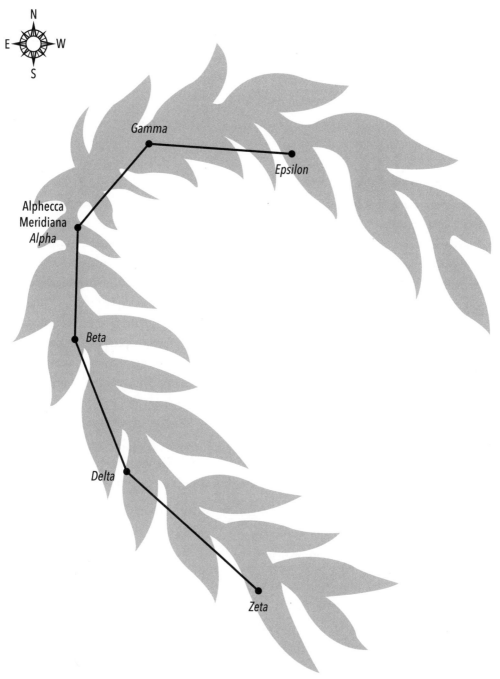

Figure 8.10. Corona Australis is the celestial heart that represents the power of love.

Official Designation: Beta Coronae Australis

Although this orange star rivals Alphecca Meridiana in brightness, it does not have a traditional name. It occupies the lower position in the trio of bright stars.

Official Designation: Gamma Coronae Australis

Gamma is a yellow-white binary star positioned at the top of the three bright stars. It does not have a traditional name.

Magical Interpretations and Uses for Corona Australis

Although Corona Australis is described as a semicircle, its star pattern resembles half of a heart. Use imagination and intention to create the other half, making it whole. Located at the core of the body, the heart is considered as the center of emotions. In many traditional cultures and religions, it is also the sacred center of being and spirit. The heart is the everlasting symbol of love, and as such, the energy of this constellation can aid in sparking love.

Prepare a red, pink, or white candle with jasmine or rose oil. Light the candle and then lay out the Corona Australis star pattern using red agate, garnet, pink jasper, or red tourmaline creating half a heart. Draw down the energy of the constellation. Use your finger to trace the other half of the heart three times as you say, filling in the blank with the other person's name: *"Corona Australis, crown like a heart; How can I give this love a start? If this relationship is to be; May it blossom for _____ and me."* End by drawing a circle with your finger around the gemstones.

Because the word "crown" is in the name of this constellation, like Corona Borealis we can consider this symbolism. Refer to chapter 4 for details on Corona Borealis.

Lupus: The Wolf/Spirit Guide

To Find: Locate the red star Antares, which marks the heart of Scorpius. Lupus is to the south and slightly west.

The wolf is a symbol of community, loyalty, protection, and spirit. It also represents freedom and the power of the wilderness as well as discipline and the power of the group.

The wolf is a spirit animal and a powerful ally for psychic and shamanic work. If a wolf presents itself during astral travel, it will be your guide and guardian. The Lupus constellation can help you call on the power of the wolf for these and other endeavors.

Ophiuchus and Serpens: The Serpent Bearer and the Serpent/Harmonic Energy

To Find: Locate Antares in Scorpius and Ophiuchus is just to the north. The triangle of
   stars to the northwest marks the head of Serpens.

These constellations can be instrumental in activating Kundalini energy, which is described as a sleeping serpent coiled around the first chakra at the base of the spine. When this energy is activated, it rises through two energy channels that weave back and forth across a central path along the spine. Working with these channels and the energy of these constellations can help us generate powerful personal energy for ritual and magic as well as healing.

Sagittarius: The Archer/Wildness of Nature

To Find: Locate Antares in Scorpius. Sagittarius is directly east of Scorpius.

Centaurs symbolize male virility, stallion energy, and sexuality in general. They epitomize sensuality and aid in developing comfortable awareness of the physical body. Although the visualization in chapter 5 was written for the summer season, it can be adapted for any time of year. Sagittarius also serves as a reminder to aim high and put our talents out for all to see. When making changes in your life, call on the energy of this constellation for aid in targeting and reaching your goals.

Scorpius: The Scorpion/Death Wielder and Protector

To Find: Locate Antares, which marks the heart of this scorpion. A line of stars that trail to the southeast delineate its tail. Scorpius is west of Sagittarius.

Although the scorpion has had negative connotations, in some cultures it was a symbol of protection. This constellation can aid in attracting good luck and repelling negativity. Scorpius also helps us embrace the dark, the place of incubation for creativity, spirituality, transformation, and clairvoyance. Let Scorpius give your efforts a boost when engaging in activities to explore deeper aspects of the self and/or develop psychic abilities.

SUMMARY

Star Magic provides us with twenty-first-century methods for drawing on the power of the constellations for ritual, magic, and other purposes. We have learned about the history of astronomy and astrology, background information about the constellations, and how to find our way among the stars throughout the seasons. We have learned that working with stars is no different from drawing down the moon or calling on other energy of the natural world.

To aid us in working with star energy, we use the palm chakras and the seven major chakras within the physical body as well as the three celestial chakras above our heads and the earth star chakra below our feet. In addition to physically sensing energy, engaging the mind with visualization is another important component in magic work. When we draw on star energy or any magical energy we become a conduit for it. We draw it in, shape it with our willpower through visualization, and then send it out along with some of our own energy. Star energy can also be released into a talisman, crystal, or something fashioned for a spell. Additionally, we can draw down star energy for use in ritual and, quite appropriately, before astrology work or astral travel. Working with the constellations carries us through the seasons, supplementing and enhancing our Wheel of the Year. As we observe and work with the cycle of energy in the green world, the stars complete the sphere of nature with a changing ceiling that complements what goes on below on the earth. As above, so below; the flow is unbroken and never ending.

As you explore and work with the constellations in either hemisphere, you may find that a few stars have more meaning for you than others. These become your special stars from which you can draw energy. Choosing one for each quarter of the year provides a guiding star for each season. Your attraction to a particular star may be a combination of factors such as its history, energy, name, or certain details about the constellation of which it is a member. Learn to find it whenever you step outside on a starry night, or if it is a less bright star, know where it is in the sky. Call on your seasonal stars to aid you in ritual, spellwork, or other endeavors. Spend time each season drawing down the energy of your special quadrant star and weave your own story around it.

In addition, you may find that a particular constellation resonates with you. If so, find a special place in your home where you can lay out its star pattern with whatever objects feel appropriate. To work with its energy more closely, lay out the pattern on your altar and gaze at it for a few minutes, and then close your eyes. Hold the image in your mind and meditate on the constellation as you draw its energy. Open yourself to receive knowledge of what this constellation means to you personally. Whenever you need to call on it, bring the image of this constellation into your mind and feel its energy surround you.

While we may follow an earth-centered spirituality, the energy of the stars envelopes our planet and holds us in the great web of the cosmos. All we need to do is reach for the stars and draw down their magic.

> *Twinkle, twinkle magic star,*
> *I see your light from afar.*
> *Up above the world so high,*
> *Shining guardian in the sky.*
> *Twinkle, twinkle celestial guide,*
> *Send your wisdom far and wide.*

Appendix A
FINDING YOUR LATITUDE

The following list provides latitude information for a number of cities around the world to help you determine if you will be able to see a particular constellation. Each constellation entry includes details about the latitudes between which they can be viewed. The equator is at latitude 0 degrees. There are 90 degrees north and south of the equator. Each degree is approximately 69 miles. The degree is subdivided into minutes, with 60 minutes in a degree. The list shows degrees and minutes of latitude. Minutes can be further subdivided; however, it is not necessary for determining whether or not a constellation can be viewed from your location.

Of course, there's an app for this. The GPS smartphone apps can show you your exact position. From your computer at home, type in your address at iTouchMap.com (www. itouchmap.com/latlong.html) for your latitude and longitude as well as a satellite view of your neighborhood.

From the North Pole to the Equator

Barrow, Alaska 71° 18′
Fairbanks, Alaska 64° 50′
Helsinki, Finland 60° 10′
Oslo, Norway 59° 54′

Inverness, Scotland 57° 28′

Liverpool, England 53° 24′

London, England 51° 30′

Calgary, Canada 51° 2′

Frankfurt am Main, Germany 50° 6′

Vancouver, Canada 49° 15′

Paris, France 48° 51′

Munich, Germany 48° 8′

Seattle, Washington 47° 36′

Quebec City, Canada 46° 48′

Geneva, Switzerland 46° 11′

Portland, Oregon 45° 31′

Montreal, Canada 45° 30′

Venice, Italy 45° 26′

Minneapolis, Minnesota 44° 59′

Toronto, Canada 43° 39′

Portland, Maine 43° 39′

Buffalo, New York 42° 53′

Boston, Massachusetts 42° 21′

Chicago, Illinois 41° 52′

New York City 40° 42′

Philadelphia, Pennsylvania 39° 57′

Baltimore, Maryland 39° 17′

St. Louis, Missouri 38° 37′

Athens, Greece 37° 59′

San Francisco, California 37° 46′

Las Vegas, Nevada 36° 10′

Tokyo, Japan 35° 41′

Oklahoma City, Oklahoma 35° 28′

Los Angeles, California 34° 3′

Phoenix, Arizona 33° 26′

Jacksonville, Florida 30° 19′

Cairo, Egypt 30° 2′

Houston, Texas 29° 45′

Miami, Florida 24° 47′

Hong Kong, China 22° 23′

Honolulu, Hawaii 21° 18′

Mexico City, Mexico 19° 25′

Bombay (Mumbai), India 19° 1′

San Juan, Puerto Rico 18° 27′

Bangkok, Thailand 13° 43′

Caracas, Venezuela 10° 30′

San Jose, Costa Rica 9° 55′

Bogota, Colombia 4° 35′

Singapore 1° 21′

FROM THE EQUATOR SOUTH

Mombasa, Kenya 4° 2′

Townsville, Australia 19° 15′

Rio de Janeiro, Brazil 22° 54′

Sao Paulo, Brazil 22° 33′

Johannesburg, South Africa 26° 12′

Brisbane, Australia 27° 28′

Santiago, Chile 33° 28′

Cape Town, South Africa 33° 55′

Buenos Aires, Argentina 34° 36′

Melbourne, Australia 37° 48′

Christchurch, New Zealand 43° 31′

Stanley, Falkland Islands UK 51° 41′

Puerto Williams, Chile 54° 55′

There are no major towns from 60° south to the south pole.

Appendix B

THE STARS AND COLOR MAGIC

We often draw on the power of color in our magic and rituals with candles and gemstones. We also use color as a way to incorporate magic into our everyday lives. Simply put, we could say that color is a tool and leave it at that; however, it can be a more powerful tool when we understand how it works. Color affects us on many levels, both conscious and subconscious. It has a profound effect on our physical, mental, emotional, and spiritual states. Color is a universal language that advertisers have used for a long time as a subtle way to spark interest in their products. We can't touch, smell, or taste color, so what is it? Think rainbow.

Colors are light waves of specific lengths. They are vibrations and a property of light. Remember the cones and rods of our eyes mentioned in chapter 1? They provide a way for our minds to perceive color; however, color is also energy.

Energy is something that we can perceive with our subtle and physical bodies. The seven major chakra energy centers within our bodies vibrate to different wavelengths of color. When they vibrate at the right level, we are in good physical and psychological health. Through the use of color, chakra energy that is out of whack can be brought back into balance. Color therapy isn't new or New Age; it dates back to the ancient Greeks and Egyptians. For our purposes in magic and ritual, the wavelengths of color reinforce and boost our intentions, allowing us to draw particular energies toward us or push them away.

In addition to the practices I have introduced for each constellation, working with the colors of their stars can amplify the power of your magical purpose. Most stars may appear white or bluish-white when viewed without binoculars or a telescope, but some stars are more colorful. Actually, the color of some of the brightest stars can be seen with the naked eye. A way to amplify a star's energy for magic is to coordinate its color with that of a candle, gemstone, or other object that you may use when laying out the star pattern of a constellation.

As we have seen, a star is not always "a" star, as it can be a double or a multiple. This means that without a telescope it may appear as a single star when it is actually more than one. These may be binary stars, two stars that revolve around each other, or a complex star system. Having more than one color for a star provides flexibility for color magic. When laying out a star pattern, you can use the color of the primary star or use multiple objects to represent each component of the star. For example, Regulus, the alpha star in Leo, has three components and each is a different color. You can use the color of Alpha-1, the primary component, which is blue-white, or you can use three objects to represent each of the components. Alternatively, you can choose the color of the Alpha-2 (orange) or the Alpha-3 (red) and use just one of these colors.

Table B.1 provides some of the basic correspondences associated with the colors we find in the stars. In the case of blue-white or yellow-white stars, use a combination of the correspondences listed.

The star color listings that follow are grouped by season. Within each season, the stars are listed according to constellation to make the information easy to find. The graphic listed with each constellation in chapters 4 through 8 includes the designation for each star. Use these star maps to lay out a constellation's pattern according to color.

Table B.1. Star Colors	
Blue	Awareness, clarity, communication, emotions, fertility, fidelity, freedom, guidance, harmony, honor, hope, intuition, knowledge, peace, prosperity, protection, purification, relationships, spirituality, transformation, wisdom Especially useful for astral travel and dream work.
Green	Abundance, balance, creativity, fertility, growth, kindness, love, luck, prosperity, renewal, success, wealth Especially useful for healing and connecting with spirits.
Lilac/ Purple	Awareness, enlightenment, happiness, inspiration, manifestation, spirituality, wisdom Especially useful for astral travel, clairvoyance, divination, and psychic work.
Orange	Abundance, adaptability, beginnings, confidence, discipline, justice, luck, power, reconciliation, strength, transformation Especially useful for dealing with change and healing.
Red	Action, courage, creativity, determination, fertility, leadership, love, loyalty, motivation, passion, power, protection, relationships, renewal, strength, willpower Especially useful for activating energy and breaking hexes.
White	Clarity, compassion, enlightenment, gratitude, happiness, harmony, hope, inspiration, peace, protection, purity, spirituality, willpower Especially useful for astral travel, divination, healing, and psychic work.
Yellow	Awareness, clarity, communication, creativity, devotion, faith, friendship, happiness, inspiration, wisdom Especially useful for astral travel, clairvoyance, dream work, and contacting spirits.

The Northern Hemisphere

Spring Star Colors by Constellation

Boötes

Alpha (Arcturus)	orange
Beta (Nekkar)	yellow
Delta	yellow-white
Epsilon	orange
Gamma	white
Kappa-1	yellow
Kappa-2	blue
Lambda	white
Rho	orange
Theta	yellow-white

Canes Venatici

Alpha-1 (Cor Caroli)	yellow
Alpha-2 (Cor Caroli)	white
Beta (Asterion/Chara)	yellow
Gamma (La Superba)	red

Centaurus

Alpha-1 (Rigil Kent)	yellow-white
Alpha-2 (Rigil Kent)	orange
Proxima	red
Beta (Agena)	blue-white
Delta	blue-white
Epsilon	blue-white
Gamma	white
Iota	white
Nu	blue-white
Theta	orange
Zeta	blue-white

Corona Borealis

Alpha-1 (Alphecca)	white
Alpha-2 (Alphecca)	yellow
Beta	yellow-white
Delta	yellow
Epsilon	orange
Gamma	blue-white
Iota	white
Theta	blue-white

Corvus

Alpha	white
Beta	yellow-white
Delta-1 (Algorab)	yellow-white
Delta-2 (Algorab)	lilac/purple
Epsilon	orange
Eta	yellow-white
Gamma (Gienah)	blue-white

Crater

Alpha (Alkes)	orange
Beta	white
Delta (Labrum)	yellow
Epsilon	orange
Eta	white
Gamma	white
Theta	blue-white
Zeta	yellow

Hydra

Alpha (Alphard)	red
Beta	blue-white
Chi-1	white
Chi-2	blue
Delta	white
Epsilon-1	yellow
Epsilon-2	yellow-white
Epsilon-3	yellow-white
Eta	blue-white
Gamma	yellow
Iota	orange
Kappa	blue-white
Lambda	orange
Mu	orange
Nu	orange
Pi	orange
Rho	white
Sigma	orange
Theta	blue-white
Upsilon-1	yellow
Zeta	yellow

Leo

Alpha-1 (Regulus)	blue-white
Alpha-2 (Regulus)	orange
Alpha-3 (Regulus)	red
Beta (Denebola)	blue
Delta (Zosma)	blue-white
Epsilon	yellow
Eta	white
Gamma-1 (Algieba)	orange
Gamma-2 (Algieba)	yellow
Mu	orange
Theta	white
Zeta	yellow-white

Libra

Alpha-1 (Zubenelgenubi)	yellow
Alpha-2 (Zubenelgenubi)	blue
Beta (Zubeneschamali)	blue-green
Gamma	orange
Sigma	red

Ursa Major

Alpha (Dubhe)	orange
Beta (Merak)	blue-white
Delta	white
Epsilon	white
Eta	blue-white
Gamma	white
Nu	orange
Omicron	yellow
Zeta	white
23 Ursae Majoris	yellow-white

Ursa Minor

Alpha (Polaris)	yellow-white
Beta	orange
Delta	white
Epsilon	yellow
Eta	yellow-white
Gamma	white
Zeta	white

Virgo

Alpha (Spica)	blue-white
Beta	yellow-white
Delta	red
Epsilon (Vindemiatrix)	yellow
Eta	white
Gamma (Porrima)	yellow-white
Mu	yellow
Tau	white
Theta	white
Zeta	white
109 Virginis	white

Summer Star Colors by Constellation

Aquila

Alpha (Altair)	yellow
Beta-1 (Alshain)	yellow
Beta-2 (Alshain)	red
Delta	yellow-white
Eta	yellow-white
Gamma (Tarazed)	orange
Lambda	blue-white
Theta	blue-white
Zeta	blue-white

Ara

Alpha	blue
Beta	orange
Delta-1	blue-white
Delta-2	yellow
Epsilon-1	orange
Eta	orange
Gamma	blue-white
Zeta	orange

Capricornus

Alpha-1 (Algiedi/Prima Giedi)	yellow
Alpha-2 (Algiedi/Secunda Giedi)	yellow
Beta-1a (Dabih/Dabih Major)	orange
Beta-1b (Dabih/Dabih Major)	blue-white
Beta-2a (Dabih/Dabih Minor)	blue-white
Beta-2b (Dabih/Dabih Minor)	white
Delta-1 (Deneb Algedi)	white
Delta-2 (Deneb Algedi)	yellow-white
Gamma	blue-white
Iota	yellow
Pi	blue-white
Psi	yellow-white
Omega	red
Theta	white
Zeta-1	yellow
Zeta-2	white
24 Capricorni	red

Cygnus

Alpha (Deneb)	blue-white
Beta-1 (Albireo)	yellow
Beta-2 (Albireo)	blue
Delta-1	blue-white
Delta-2	yellow-white
Delta-3	orange
Epsilon (Gienah)	orange
Eta	orange
Gamma	yellow-white
Theta	yellow-white
Zeta	yellow

Draco

Alpha (Thuban)	white
Beta (Rastaban)	yellow
Chi	yellow-white
Delta	yellow
Epsilon	yellow
Eta	yellow
Gamma (Eltanin)	orange
Iota	orange
Kappa	blue
Lambda	red
Theta	yellow-white
Xi	orange
Zeta	blue-white

Hercules

Alpha-1a (Rasalgethi)	orange
Alpha-1b (Rasalgethi)	blue-green
Alpha-2a (Rasalgethi)	yellow
Alpha-2b (Rasalgethi)	yellow-white
Beta (Kornephoros)	yellow
Delta	white
Epsilon	white
Eta	yellow
Iota	blue-white
Kappa-1	yellow
Kappa-2	orange
Mu	yellow
Omega	blue-white
Omicron	blue-white
Pi	orange
Sigma	white
Tau	blue-white
Theta	orange
Zeta	yellow-white

Lupus

Alpha (Kakkab)	blue-white
Beta	blue-white
Delta	blue-white
Epsilon -1	blue-white
Epsilon-2	blue-white
Epsilon-3	white
Eta	blue-white
Gamma (Thusia)	blue-white
Phi-1	orange
Zeta	yellow

Lyra

Alpha (Vega)	blue-white
Beta (Sheliak)	blue-white
Delta-1a	blue-white
Delta-1b	orange
Delta-2	red
Gamma (Sulafat)	blue-white
Zeta	white

Ophiuchus

Alpha-1 (Rasalhague)	white
Alpha-2 (Rasalhague)	orange
Beta (Cebalrai)	orange
Eta	white
Gamma	white
Kappa	orange
Phi	yellow
Theta	blue-white
Zeta	blue

Sagittarius

Alpha (Rukbat)	blue
Beta-1 (Arkab Prior)	blue
Beta-2 (Arkab Posterior)	yellow-white
Delta	orange
Epsilon-1 (Kaus Australis)	blue
Epsilon-2 (Kaus Australis)	orange
Gamma (Alnasl)	orange
Lambda	orange
Omega	yellow
Phi	blue-white
Psi	white
Sigma	blue-white
Tau	orange
Theta-1	blue
Theta-2	white
Zeta	white

Scorpius

Alpha (Antares)	red
Beta (Acrab)	blue-white
Delta	blue-white
Epsilon	orange
Eta	yellow-white
Iota-1	yellow-white
Lambda (Shaula)	blue-white
Pi	blue-white
Tau	blue-white
Theta	yellow
Zeta-1	blue-white
Zeta-2	orange

Serpens Caput

Alpha (Unukalhai)	orange
Beta	white
Delta	yellow-white
Gamma	white
Kappa	red
Mu	white

Serpens Cauda

Eta	orange
Theta (Alya)	white
Xi	yellow-white

Autumn Star Colors by Constellation

Andromeda

Alpha (Alpheratz)	blue
Beta (Mirach)	red
Delta-1	orange
Delta-2	yellow
Gamma-1 (Almach)	orange
Gamma-2 (Almach)	blue
Gamma-3 (Almach)	white
Mu	white
Pi	blue
Zeta	orange
51 Andromedae	orange

Aquarius

Alpha (Sadalmelik)	yellow
Beta (Sadalsuud)	yellow
Delta (Skat)	yellow
Eta	blue
Gamma	white
Lambda	red
Nu	yellow
Tau-2	red
Zeta	yellow-white
88 Aquarii	orange

Aries

Alpha (Hamal)	orange
Beta-1 (Sheratan)	white
Beta-2 (Sheratan)	yellow
Delta	orange
Epsilon	white
Gamma-1 (Mesarthim)	white
Gamma-2 (Mesarthim)	white
Gamma-3 (Mesarthim)	orange
Zeta	white

Cassiopeia

Alpha (Schedar)	orange
Beta (Caph)	yellow-white
Delta (Ruchbah)	blue-white
Epsilon	blue-white
Gamma	blue

Cepheus

Alpha (Alderamin)	white
Beta-1 (Alfirk)	blue
Beta-2 (Alfirk)	white
Beta-3 (Alfirk)	white
Gamma (Errai)	orange
Iota	orange
Mu (the Garnet Star)	red
Zeta	orange

Cetus

Alpha (Menkar)	red
Beta (Deneb Kaitos)	orange
Delta	blue
Eta	orange
Gamma	white
Iota	orange
Lambda	blue-white
Mu	yellow-white
Omicron-1 (Mira)	red
Omicron-2 (Mira)	white
Theta	orange
Xi-1	white
Xi-2	blue-white
Zeta (Baten Kaitos)	orange

Delphinus

Alpha (Sualocin)	blue-white
Beta (Rotanev)	yellow-white
Delta	white
Epsilon (Deneb al Dulfin)	white
Gamma-1	yellow-white
Gamma-2	orange

Equuleus

Alpha (Kitalpha)	yellow
Beta	white
Delta-1	yellow
Delta-2	yellow-white
Gamma	white

Pegasus

Alpha (Markab)	blue-white
Beta (Scheat)	red
Gamma (Algenib)	blue-white
Epsilon (Enif)	orange
Kappa	yellow-white
Pi-2	yellow-white
Theta	white

Pisces

Alpha (Alrescha)	white
Chi	orange
Delta	orange
Eta (Kullat Nunu)	yellow
Gamma	yellow
Iota	yellow-white
Kappa	white
Lambda	white
Mu	orange
Omicron	yellow
Tau	orange
Theta	orange
Upsilon	white
Zeta	white
7 Piscum	orange

Winter Star Colors by Constellation

Auriga

Alpha-1 (Capella)	yellow
Alpha-2 (Capella)	red
Beta (Menkalinan)	blue-white
Epsilon-1	yellow-white
Epsilon-2	blue-white
Eta	blue-white
Gamma	blue
Iota	orange
Theta-1	white
Theta-2	yellow
Zeta-1	red
Zeta-2	blue-white

Cancer

Alpha (Acubens)	white
Beta (Al Tarf)	orange
Delta	orange
Eta	orange
Gamma	white
Iota-1	yellow
Iota-2	white
Mu-1	red
Mu-2	yellow
Theta	orange

Canis Major

Alpha (Sirius)	white
Beta	blue-white
Delta	white
Epsilon (Adhara)	blue-white
Eta	blue
Gamma	blue-white
Iota	blue-white
Theta	orange

Canis Minor

Alpha (Procyon)	white
Beta	blue-white
Epsilon	yellow
Gamma	orange
Zeta	blue-white

Eridanus

Alpha-1 (Achernar)	blue
Alpha-2 (Achernar)	white
Beta (Cursa)	white
Delta	orange
Eta	orange
Gamma (Zaurak)	yellow
Lambda	blue-white
Mu	blue-white
Phi	blue-white
Tau-1	white
Tau-3	white
Tau-5	blue
Tau-8	blue
Theta	white
Upsilon-2	yellow
Upsilon-4	blue-white
Xi	white
Zeta	white

Gemini

Alpha-1 (Castor)	white
Alpha-2 (Castor)	white
Alpha-3 (Castor)	red
Beta (Pollux)	orange
Delta	yellow-white
Epsilon	yellow
Gamma	white
Iota	yellow
Kappa	yellow
Lambda-1	white
Lambda-2	yellow
Lambda-3	orange
Mu	red
Nu	blue-white
Theta	white
Xi	yellow-white
Zeta	yellow

Lepus

Alpha (Arneb)	yellow-white
Beta (Nihal)	yellow
Eta	white
Epsilon	orange
Gamma	yellow-white
Kappa	blue-white
Lambda	blue-white
Mu	blue-white
Theta	white
Zeta	white

Monoceros

Alpha Monocerotis	orange
Beta Monocerotis	blue-white
Delta Monocerotis	white
Gamma Monocerotis	orange
Zeta Monocerotis	yellow

Orion

Alpha (Betelgeuse)	red
Beta (Rigel)	red
Chi-1	yellow
Delta-1	blue-white
Delta-2	blue
Epsilon	blue
Gamma (Bellatrix)	blue
Kappa	blue
Lambda	blue
Nu	blue
Pi-1	white
Pi-2	white
Pi-3	white
Xi	blue
Zeta	blue

Perseus

Alpha (Mirfak)	yellow-white
Beta-1 (Algol)	blue-white
Beta-2 (Algol)	orange
Beta-3 (Algol)	white
Chi	blue-white
Epsilon-1	blue-white
Epsilon-2	white
Epsilon-3	orange
Eta	orange
Gamma-1	yellow
Gamma-2	white
Kappa	orange
Lambda	blue
Mu	yellow

The Pleiades

Eta-1 (Alcyone)	blue-white
Eta-2 (Alcyone)	white
Eta-3 (Alcyone)	white
Eta-4 (Alcyone)	yellow-white
16 Tauri (Celaeno)	blue-white
17 Tauri (Electra)	blue-white
19 Tauri-1 (Taygeta)	blue-white
19 Tauri-2 (Taygeta)	blue-white
19 Tauri-3 (Taygeta)	white
20 Tauri (Maia)	blue-white
21 Tauri (Asterope)	blue-white
22 Tauri (Asterope)	blue-white
23 Tauri (Merope)	blue-white

Taurus

Alpha (Aldebaran)	orange
Beta (Elnath)	blue
Epsilon	orange
Eta-1 (Alcyone)	blue-white
Eta-2 (Alcyone)	white
Eta-3 (Alcyone)	white
Eta-4 (Alcyone)	yellow-white
Gamma	orange
Kappa	white
Zeta	blue-white

THE SOUTHERN HEMISPHERE

Star Colors by Constellation

Corona Australis

Alpha (Alphecca Meridiana)	white
Beta	orange
Delta	orange
Epsilon	white
Gamma	yellow-white
Zeta	blue-white

Crux

Alpha (Acrux)	blue-white
Beta (Mimosa)	blue-white
Delta Crucis	blue-white
Gamma (Gacrux)	red

Grus

Alpha (Al Nair)	blue-white
Beta (Al Dhanab)	red
Delta-1a	yellow
Delta-1b	orange
Delta-2	red
Epsilon	white
Gamma	blue
Lambda	orange
Rho	orange
Zeta	orange

Hydrus

Alpha	yellow-white
Beta	yellow
Delta	white
Eta-1	blue-white
Eta-2	yellow
Gamma	red
Nu	orange
Zeta	white

Phoenix

Alpha (Ankaa)	orange
Beta	yellow
Epsilon	orange
Eta	white
Gamma	red
Iota	white
Kappa	white

Piscis Austrinus

Alpha (Fomalhaut)	white
Beta	white
Delta	yellow
Epsilon	blue-white
Gamma	white
Iota	white
Lambda	blue
Mu	white
Theta	white

Appendix C

THE FIXED STARS
OF MEDIEVAL MAGIC

While we know that twinkling is a way to tell the difference between a star and a planet, people in ancient times and up through the Middle Ages did not. They did, however, notice a difference in behavior and made intelligent distinctions. In ancient Egypt, stars were called imperishable stars and the planets were called the stars that never rested.[33] In medieval Europe, the stars and planets were called fixed stars and wandering stars, respectively. Fixed stars rose and set as did the sun, but they seemed to stay in the same pattern in relation to other stars. The planets were called wandering stars because their positions changed within a shorter period of time. They also seemed to move independently of other stars.

The fifteen stars noted by Agrippa were considered particularly powerful for magic by medieval astrologers in Europe and the Middle East. These can be considered as the high-profile stars that Agrippa called the Behenian fixed stars. This name has caused some confusion. According to some sources, the word "behenian" comes from Arabic *bahman*, meaning "root." However, according to author and translator James Freake, Agrippa used the term behenian "as a synonym for Arabian." [34] The word "behen" was derived from the Arabic *bahman*, which is a particular kind of root. Native to the Euphrates Valley, *Withania somnifera* was known as *Bahman Root* in parts of the Middle East and is still used in

Ayurvedic medicine today. In medieval times it was used as a cure for various ailments and to protect against evil. As a result, these stars were considered important medicinally and magically.

Through the Middle Ages and Renaissance, the use of fixed stars for high magic and astrology was common practice, but because some stars have rather fatal or negative associations the use of fixed stars in astrology gradually fell out of favor. However, modern astrologers have rediscovered these stars as a source of knowledge and have been using them as a way to add information to readings. Today, astrologers point out that the negative aspects ascribed to some stars serve as warnings and point out aspects of life of which one may need to be particularly mindful.

From early times, astrologers equated the attributes and energy of the planets with the constellations and individual stars. These planets consisted of the traditional five (Mercury, Venus, Mars, Jupiter, and Saturn) as well as the sun and moon, which were often referred to as luminaries. Since then, the outer planets have been discovered and modern astrology, quite naturally, includes them. The following table provides a brief overview of these energies and attributes.

Table C.1. Planetary Attributes of Astrology	
Sun	Personal identity, sense of who you are, the self, the ego
Moon	Inner identity, ideals, emotions, what you seek, the soul
Mercury	The intellect, ideas, communications, your mental picture of the world
Venus	Creative expression, beauty, relationships, social skills, love
Mars	Energy, action, initiative, motivation, sexual drive
Jupiter	Expansion/learning and integration, growth, luck, achievement
Saturn	Learning life's lessons, responsibility, building a foundation and structure for life
Uranus	Independence, different approaches, change
Neptune	Vision, creativity, illusion
Pluto	Transformation, regeneration, rebirth

According to Agrippa, the Behenian fixed stars were a source of power for the planets. One way that Agrippa assigned planetary equivalents to stars was by color. However, this system did not account for all the various star colors, which outnumber the colors of planets. It also did not take into consideration the fact that many are actually double or multiple stars, which in most cases he would not have known without a high-powered telescope.

Working with what he knew, Agrippa correlated the energy of particular plants and gemstones commonly used in ritual and magic with the energy of stars to draw their power. You may want to try these plants and gemstones if you work with these stars. Likewise, the sigils associated with fixed stars were used as talismans to attract and hold the energy of the stars. Although Agrippa presented the sigils in his work, their origin may date back to an Egyptian source.

Just as we learned to meditate and hold the image of a constellation in our minds, the sigils can be used in the same way. They can also be carved into candles or painted on gemstones and other items that we want to use as talismans. When laying out a star pattern, you may want to substitute a star's sigil instead of the type of object you are using for the other stars in the constellation. In addition, you may come up with your own sigils for these stars or any others with which you work.

Agrippa's Fifteen Powerful Stars

Aldebaran

Official Designation: Alpha Tauri

Constellation: Taurus

Gemstones: Garnet, ruby

Plants: Milk thistle, woodruff

Aldebaran served as one of the four royal stars of Persia as the Guardian of the East. Ptolemy equated it with Mars; however, later astrologers equated it with Mars and Venus as well as Mercury, Mars, and Jupiter. Aldebaran is associated with honor, intelligence, eloquence, steadfastness, courage, honesty, and success.

Algol

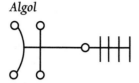

Official Designation: Beta Persei

Constellation: Perseus

Gemstone: Diamond

Plants: Black hellebore, mugwort

Ptolemy equated this star with Saturn and Jupiter. In ancient times, it was considered one of the most evil stars and linked with demons. The reason for this may be because it was associated with the independent and strong-willed Lilith. To some ancient people, female power and sexual energy was considered an abomination. This star is now associated with strength, intense passion, and the forces of the natural world.

Alphecca

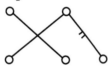

Official Designation: Alpha Coronae Borealis

Constellation: Corona Borealis

Gemstone: Topaz

Plants: Rosemary, trefoil, ivy

Ptolemy equated this star with Venus and Mercury; Agrippa with Venus and Mars. Later astrologers equated it with Mars and Mercury. This star is associated with love, honor, artistic skills, quiet achievement, and a change in social status that is earned.

Antares

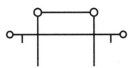

Official Designation: Alpha Scorpii

Constellation: Scorpius

Gemstones: Amethyst, sardonyx

Plants: Saffron, snakeroot

Antares was one of the four royal stars of Persia and called the Guardian of the West. Ptolemy equated it with Mars and Jupiter. While in the past it was associated with evil and destructiveness, it is now considered to toughen a person for dealing with difficult issues. It also indicates that one should be mindful of the potential for self-destruction. Antares is an aid in driving away evil spirits and can be used for defense and protection.

Arcturus

Official Designation: Alpha Boötis

Constellation: Boötes

Gemstone: Jasper

Plant: Plantain

According to Ptolemy, this star is equated with Mars and Jupiter. It is associated with protection and guidance (a night-watcher of sorts) and success in the arts. Additionally, it is linked with learning, teaching, leading, and exploring.

Capella

Official Designation: Alpha Aurigae

Constellation: Auriga

Gemstone: Sapphire

Plants: Horehound, mint, mugwort, thyme

Ptolemy equated this star with Mars and the moon; Agrippa with Jupiter and Saturn. It is associated with honors, public position, and wealth. Capella is also linked with ambition but warns that a person needs to be mindful that he or she does not let it get out of hand.

Deneb Algedi

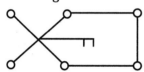

Official Designation: Delta Capricorni

Constellation: Capricornus

Gemstone: Chalcedony

Plants: Mandrake, marjoram, mugwort

While Ptolemy equated this star with Jupiter and Saturn, Agrippa noted that it was more like Saturn and Mercury. It is associated with wisdom, integrity, and justice. In the past, Deneb Algedi was linked with opposites: sorrow and happiness, life and death. Today's interpretation associates it with the fullness of life and the importance of balance.

Gienah / Algorab

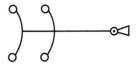

Official Designation: Gamma Corvi and Delta Corvi

Constellation: Corvus

Gemstone: Onyx

Plants: Burdock, henbane, comfrey

Ptolemy equated this star with Saturn and Mars. In medieval magic, it was used to drive away evil spirits.

The Pleiades

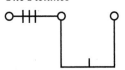

Official Designation: M45

Constellation: Taurus

Gemstone: Quartz crystal

Plants: Frankincense, fennel

According to Ptolemy, the Pleiades were equated with Mars and the moon. They are associated with love, eminence, and seeking inner knowledge. These stars are instrumental for communication with spirits and engendering peaceful energy.

Polaris

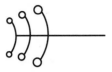

Official Designation: Alpha Ursae Minoris

Constellation: Ursa Minor

Gemstone: Lodestone

Plants: Chicory, mugwort, periwinkle

Ptolemy equated this star with Saturn and a little bit with Venus. Agrippa thought that Venus and the moon were more appropriate. In medieval magic, it was used for protection against spells. Polaris is associated with having a sense of where you want to go in life and having a mission.

Procyon

Official Designation: Alpha Canis Minoris

Constellation: Canis Minor

Gemstone: Agate

Plants: Buttercup, marigold, pennyroyal

Ptolemy equated Procyon mostly with Mercury and a little with Mars. In the past it was linked with dangers. Now it is associated with power, good health, wealth, and fame. The caution with this star is to be mindful of fortune and success because it can slip away. Also, don't look for quick profit or rapid success as they may be short lived.

Regulus

Official Designation: Alpha Leonis

Constellation: Leo

Gemstones: Garnet, granite

Plant: Mugwort

Regulus was one of the royal stars of Persia and known as the Guardian of the North. Ptolemy equated it with Jupiter and Mars. It is associated with power, success, and strength. The warning with Regulus is that once these attributes are attained they should never be used for revenge; otherwise all that was gained will be lost.

Sirius

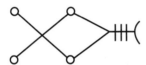

Official Designation: Alpha Canis Majoris

Constellation: Canis Major

Gemstone: Beryl

Plants: Juniper, dragonwort

Ptolemy equated Sirius with Jupiter and a little with Mars. It is associated with communication, marital peace, passion, faithfulness, and wealth. Sirius is also considered a guardian.

Spica

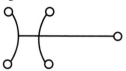

Official Designation: Alpha Virginis

Constellation: Virgo

Gemstone: Emerald

Plants: Sage, clover

Ptolemy equated Spica with Venus and Mars; Agrippa with Venus and Mercury. This star is associated with protection, abundance, psychic abilities, knowledge, and insight.

Vega

Official Designation: Alpha Lyrae

Constellation: Lyra

Gemstone: Chrysolite

Plant: Winter savory

Sometimes called Wega, Ptolemy equated this star with Mercury and Venus. It is associated with artistic talents, social awareness, and magic as well as hopefulness and idealism. It is often used to avert fear.

METEOR SHOWERS

Like stargazing, watching for meteors is best on nights when the moon is not full. Also, like working with constellations, it is not necessary to see the meteor showers to work with their energy; however, it is important to know when they will occur. The following list provides the general range of dates for major meteor shower activity. The exact dates and the particular nights that these showers reach their peaks can vary from year to year. One of the best websites for details on meteor showers is the American Meteor Society (www.amsmeteors.org). Also, there are meteor shower calendar apps for smartphones.

Best Seen in the Northern Hemisphere:

January 1–10, Quadrantids

April 16–25, Lyrids

July 11–August 10, Alpha Capricornids*

July 13–August 26, Perseids

October 6–10, Draconids

October 4–November 14, Orionids

October 19–December 10, Northern Taurids

November 5–30, Leonids

December 4–16, Geminids

* Seen equally well in both hemispheres.

Best Seen in the Southern Hemisphere:

January 28–February 21, Alpha Centaurids

February 25–March 22, Gamma Normids

April 19–May 26, Eta Aquariids

July 21–August 23, Delta Aquariids

September 7–November 19, Southern Taurids

November 28–December 9, Phoenicids

December 17–23, Ursids

COMPLETE LIST OF CONSTELLATIONS

Following is a complete list of the eighty-eight modern constellations. The constellation names in italic were covered in this book.

Andromeda the Princess
Antlia the Air Pump
Apus the Bird of Paradise
Aquarius the Water Bearer
Aquila the Eagle
Ara the Altar
Aries the Ram
Auriga the Charioteer
Boötes the Plowman
Caelum the Sculptor's Chisel
Camelopardalis the Giraffe
Cancer the Crab
Canes Venatici the Hunting Dogs
Canis Major the Great Dog

Canis Minor the Little Dog

Capricornus the Sea Goat

Carina the Keel [of Argo Navis—extinct]

Cassiopeia the Queen

Centaurus the Centaur

Cepheus the King

Cetus the Whale

Chamaeleon the Chameleon

Circinus the Compass

Columba the Dove

Coma Berenices the Hair of Berenice

Corona Australis the Southern Crown

Corona Borealis the Northern Crown

Corvus the Crow

Crater the Cup

Crux the Southern Cross

Cygnus the Swan

Delphinus the Dolphin

Dorado the Swordfish

Draco the Dragon

Equuleus the Colt

Eridanus the River

Fornax the Furnace

Gemini the Twins

Grus the Crane

Hercules the Strongman

Horologium the Clock

Hydra the Water Snake

Hydrus the Southern Water Snake

Indus the American Indian

Lacerta the Lizard

Leo the Lion

Leo Minor the Lion Cub

Lepus the Hare

Libra the Scales

Lupus the Wolf

Lynx the Lynx

Lyra the Harp

Mensa the Table Mountain

Microscopium the Microscope

Monoceros the Unicorn

Musca the Fly

Norma the Carpenter's Level

Octans the Octant

Ophiuchus the Serpent Bearer

Orion the Hunter

Pavo the Peacock

Pegasus the Winged Horse

Perseus the Hero

Phoenix the Phoenix

Pictor the Easel

Pisces the Fish

Piscis Austrinus the Southern Fish

Puppis the Stern [of Argo Navis—extinct]

Pyxis the Mariner's Compass [of Argo Navis—extinct]

Reticulum the Net

Sagitta the Arrow

Sagittarius the Archer

Scorpius the Scorpion

Sculptor the Sculptor's Workshop

Scutum the Shield

Serpens the Serpent

Sextans the Sextant

Taurus the Bull

Telescopium the Telescope

Triangulum the Triangle

Triangulum Australe the Southern Triangle

Tucana the Toucan

Ursa Major the Great Bear
Ursa Minor the Little Bear
Vela the Sails [of Argo Navis—extinct]
Virgo the Virgin
Volans the Flying Fish
Vulpecula the Fox

GLOSSARY

asterism: An easily discernible pattern of stars not officially recognized as a constellation.

astral body: The outer aspect of the body that operates as a vehicle of consciousness and allows us to function on the astral plane.

astral plane: A realm or level of consciousness where magical energies are marshaled for manifestation on the physical plane.

astronomical twilight: The third stage of evening twilight that occurs when the horizon is no longer visible.

aura: A subtle energy field that surrounds all living things.

averted vision: A method for seeing a dim sky object without looking directly at it.

Bayer designation: A naming convention for stars in a constellation that uses Greek letters to rank their brightness. It was created by German astronomer Johann Bayer, who was the first to map the entire sky.

Behenian fixed stars: Fifteen stars defined by Agrippa as particularly powerful for magic.

binary star: A star that actually consists of two stars orbiting each other.

celestial equator: Earth's equator projected out to the sky, dividing it into north and south.

celestial pole: Points on the celestial sphere above Earth's North and South Poles.

celestial sphere: The imaginary sphere or dome around Earth that is useful in understanding the movements of celestial objects.

chakra: A center or point of energy within or near the physical body.

circumpolar: A constellation or star that appears to circle one of the celestial poles.

civil twilight: The first stage of evening twilight, when only the brightest stars can be seen.

cones and rods: The two types of light-detecting cells in our eyes. Cones are highly concentrated at the center of the eyes while rods are off to the sides.

constellation: One of eighty-eight official, defined areas of the sky that includes a recognized star pattern. These were created by the International Astronomical Union in 1922.

declination: The north/south position of an object on the celestial sphere.

double star: A star that is actually two when viewed with binoculars or a telescope. Unlike a binary star, these do not orbit each other.

eclipsing binary: A type of binary star where one component eclipses the other, producing a change in overall brightness.

ecliptic: The band or path of the sun around the celestial sphere.

equinox: One of two days of the year when the sun crosses the equator. It is the point at which the ecliptic and equator intersect on the celestial sphere. The spring or vernal equinox occurs around March 20 and the autumn equinox around September 22. These are celebrated as Ostara and Mabon, respectively.

Flamsteed numbers: A system in use today that numbers stars from west to east in each constellation without regard to their brightness. This system was created by English astronomer John Flamsteed.

full darkness: This occurs after the three stages of evening twilight, when faint stars can be seen.

horizon: A boundary where the sky meets the earth or ocean.

light pollution: The illumination of the night sky by artificial light that inhibits observation of celestial objects.

lucid star: A star that can be seen with the naked eye.

magnitude system: A method employed by astronomers to measure the brightness of stars.

meridian: An imaginary line that runs from north to south and passes through your zenith, dividing the sky into east and west.

Messier objects: Nebulae, clusters, and galaxies catalogued by French astronomer Charles Messier. Messier objects are designated by the letter *M* followed by a number.

meteor shower: Bright streaks of light that appear when meteors burn up while traveling through Earth's atmosphere.

Milky Way: The great spiral galaxy that contains our solar system, which is located on one of the smaller spurs called the Orion Arm.

multiple star: A star system consisting of three or more stars that may not be distinguishable without binoculars or a telescope.

nautical twilight: The second stage of evening twilight, when the horizon is still visible and the bright stars used for navigation appear.

nebula: An area of interstellar gas and/or dust. The word is Latin and means "cloud."

open cluster: A group of stars consisting of a few dozen or several thousand stars. Also called galactic star clusters, these have more distance between stars than other types of clusters.

planisphere: A star map with a rotating disk that can be set to show the constellations at any given day or time during the year.

precession: The scientific name for Earth's wobble, which is caused by gravitational pull of the sun and moon. This movement causes Earth's rotational axis to change, thus shifting the celestial poles and the equinoxes. It is also called the precession of the equinoxes.

right ascension: The east/west position of a celestial body.

royal stars: Four stars used as calendar markers by the Babylonians that they designated as the guardians of the cardinal directions.

Sabianism: Star worship that was part of the pre-Islamic religion of the Harranians of Mesopotamia.

solstice: One of two days in the year when the sun reaches its maximum northern or southern point. The summer solstice occurs around June 21 and the winter solstice around December 21. These are celebrated as Litha and Yule, respectively.

spectroscopic binary star: Two stars orbiting each other that are too close to be distinguished as separate without a high-powered telescope.

Tropic of Cancer: The northern latitude where the sun is directly overhead at noon on the summer solstice.

Tropic of Capricorn: The southern latitude where the sun is directly overhead at noon on the winter solstice.

twilight: The time between sunset and full darkness that occurs in three stages: civil, nautical, and astronomical. These three stages occur in reverse order at sunrise.

variable star: A star that repeatedly changes brightness.

zenith: The point directly overhead on the celestial sphere.

zodiac: Constellations that provide a backdrop to the sun along the circle of the ecliptic. Traditionally there are twelve; however, in the two thousand years since its inception, Ophiuchus now falls on the ecliptic, making it the thirteenth constellation of the zodiac.

BIBLIOGRAPHY

Agrippa, Heinrich Cornelius. *Three Books of Occult Philosophy*. Translated by James Freake. Edited by Donald Tyson. St. Paul, MN: Llewellyn Worldwide, 2004.

Alessi, Justine A., and M. E. McMillan. *Rebirth of the Oracle: The Tarot for the Modern World*. Hunstville, AR: Ozark Mountain Publishers, 2005.

Allen, Richard H. *Star Names: Their Lore and Meaning*. Mineola, NY: Dover Publications, Inc., 1963.

Andrews, Munya. *The Seven Sisters of the Pleiades: Stories from Around the World*. North Melbourne, Australia: Spinifex Press Pty Ltd., 2004.

Andrews, Ted. *Animal Speak: The Spiritual & Magical Powers of Creatures Great & Small*. St. Paul, MN: Llewellyn Publications, 2002.

Anthon, Charles. *A Latin-English and English-Latin Dictionary*. New York: Harper & Brothers Publishers, 1853.

Aratus. *Phaenomena*. Translated by Aaron Poochigian. Baltimore: The Johns Hopkins University Press, 2010.

Artress, Lauren. *Walking a Sacred Path: Rediscovering the Labyrinth as a Spiritual Tool*. New York: Riverhead Books, 1995.

Bagnal, Philip M. *The Star Atlas Companion: What You Need to Know about the Constellations*. New York: Springer Science and Business Media, 2012.

Bakich, Michael E. *The Cambridge Guide to the Constellations*. New York: Cambridge University Press, 1995.

——— . *1,001 Celestial Wonders to See Before You Die*. New York: Springer, 2010.

Barth, Edna. *Shamrocks, Harps, and Shillelaghs*. New York: Clarion Books, 2001.

Baur, Jaroslav, and Vladimir Bouška. *A Guide in Color to Precious & Semiprecious Stones*. Secaucus, NJ: Chartwell Books, Inc., 1989.

Becker, Udo. *The Continuum Encyclopedia of Symbols*. New York: The Continuum International Publishing Group, Inc., 2000.

Berg, Ruth. *The Secret Is in the Rainbow: Aura Interrelationships*. New York: Weiser Books, 1989.

Berry, Richard. *Discover the Stars*. New York: Harmony Books, 1987.

Bonnefoy, Yves, and Wendy Doniger. *Roman and European Mythologies*. Chicago: University of Chicago Press, 1992.

Brady, Bernadette. *Brady's Book of Fixed Stars*. York Beach, ME: Red Wheel/Weiser LLC, 1998.

Burnham, Robert, Jr. *Burnham's Celestial Handbook: An Observer's Guide to the Universe, Volume 3*. Mineola, NY: Dover Publications, 1978.

Burritt, Elijah Hinsdale, and Henry Whitall. *The Geography of the Heavens*. Boston: Allen & Ticknor, 1883.

Busenbark, Ernest. *Symbols, Sex and the Stars*. San Diego: The Book Tree, 2003.

Cirlot, Juan Eduardo. *A Dictionary of Symbols*. Mineola, NY: Dover Publications, 2002.

Condos, Theony, ed. *Star Myths of the Greeks and Romans*. Grand Rapids, MI: Phanes Press, 1997.

Conway, D. J. *Celtic Magic*. St. Paul, MN: Llewellyn Publications, 2002.

——— . *Magickal, Mystical Creatures: Invite Their Powers into Your Life*. St. Paul, MN: Llewellyn Publications, 2001.

Cooper, George H. *Ancient Britain: The Cradle of Civilization*. San Jose, CA: Hillis-Murgotten Co., 1921.

Curran, Bob, and Ian Daniels. *Walking with the Green Man*. Franklin Lakes, NJ: The Career Press, 2007.

Davidson, H. R. Ellis. *Myths and Symbols in Pagan Europe: Early Scandinavian and Celtic Religions*. Syracuse, NY: Syracuse University Press, 1988.

Dickinson, Terence. *NightWatch: A Practical Guide to Viewing the Universe*. Buffalo, NY: Firefly Books, 2003.

Eliade, Mircea. *Cosmos and History: The Myth of the Eternal Return*. Translated by Willard R. Trask. New York: Harper Torchbooks, 1954.

Ellis, Peter Berresford. *Early Irish Astrology*. Dublin: *Réalta* (vol 3. n. 3, 1996), the journal of The Irish Astrological Association.

Evans, James. *The History and Practice of Ancient Astronomy*. New York: Oxford University Press, 1998.

Falkner, David E. *The Mythology of the Night Sky*. New York: Springer Science & Business Media, 2011.

Fontana, David. *The Secret Language of Symbols*. San Francisco: Chronicle Books, 2003.

Franklin, Anna. *The Celtic Animal Oracle*. London: Vega Books, 2003.

Gall, James. *An Easy Guide to the Constellations*. New York: G. P. Putnam's Sons, 1910.

Gallant, Roy A. *The Constellations: How They Came to Be*. Cincinnati: Four Winds Press, 1979.

Garfinkle, Robert A. *Star-Hopping: Your Visa to Viewing the Universe*. New York: Cambridge University Press, 1994.

Gimbutas, Marija. *The Language of the Goddess*. New York: HarperCollins Publishers, 1991.

Grant, Michael, and John Hazel. *Who's Who in Classical Mythology*. New York: Routledge, 2002.

Green, Miranda. *Animals in Celtic Life and Myth*. New York: Routledge, 1992.

———. *Celtic Myth*. Austin, TX: University of Texas Press, 1998.

———. *The Celtic World*. New York: Routledge, 1996.

Green, Tamara M. *The City of the Moon God*. Leiden, The Netherlands: E. J. Brill, 1992.

Greer, John Michael. *The New Encyclopedia of the Occult*. St. Paul, MN: Llewellyn Publications, 2003.

Hall, Judy. *The Crystal Bible*. Cincinnati: Walking Stick Press, 2003.

———. *The Encyclopedia of Crystals*. Beverly, MA: Fair Winds Press, 2006.

Hanon, Geraldine Hatch. *Sacred Space: A Feminist Vision of Astrology*. Ithaca, NY: Firebrand Books, 1990.

Houck, C. M. *The Celestial Scriptures: Keys to the Suppressed Wisdom of the Ancients*. Lincoln, NE: iUniverse, Inc., 2002.

James, Matthew. *The Original Prophecy*. Indianapolis, IN: Dog Ear Publishing, 2010.

Kaler, James B. *The Ever-Changing Sky: A Guide to the Celestial Sphere*. New York: Cambridge University Press, 2002.

———. *The Little Book of Stars*. New York: Copernicus Books, 2001.

Kapoor, L. D. *Handbook of Ayurvedic Medicinal Plants*. Boca Raton, FL: CRC Press, LLC, 2001.

Keith, A. M. *The Play of Fictions: Studies in Ovid's Metamorphoses, Book 2*. Ann Arbor, MI: University of Michigan Press, 1992.

Kent, April Elliott. *The Essential Guide to Practical Astrology*. New York: Alpha Books, 2011.

Kieckhefer, Richard. *Magic in the Middle Ages*. New York: Cambridge University Press, 2000.

Leeming, David, and Jake Page. *Goddess: Myths of the Female Divine*. New York: Oxford University Press, 1994.

Lewis, James R. *The Astrology Book: The Encyclopedia of Heavenly Influences*. Canton, MI: Visible Ink Press, 2003.

Lilly, Sue, and Simon Lilly. *Healing with Crystals and Chakra Energies*. London: Hermes House, 2004.

Lindberg, David C., ed. *Science in the Middle Ages*. Chicago: University of Chicago Press, 1980.

Mackillop, James. *The Oxford Dictionary of Celtic Mythology*. New York: Oxford University Press, 1998.

MacLeod, Sharon Paice. *Celtic Myth and Religion: A Study of Traditional Belief*. Jefferson, NC: McFarland & Company, 2012.

Mammana, Dennis L. *The Night Sky: An Observer's Guide*. New York: Mallard Press, 1993.

Matthews, Caitlin. *Mabon and the Guardians of Celtic Britain*. Rochester, VT: Inner Traditions International, 2002.

McColl, R. W. *Encyclopedia of World Geography, Volume 1*. New York: Facts on File, Inc., 2005.

McColman, Carl, and Kathryn Hinds. *Magic of the Celtic Gods and Goddesses*. Franklin Lakes, NJ: The Career Press, 2005.

McIntosh, Jane. *Handbook to Life in Prehistoric Europe*. New York: Oxford University Press, 2006.

Mercier, Patricia. *The Chakra Bible*. New York: Sterling Publishing, 2007.

Mitchell, John. *Secrets of the Stones: New Revelations of Astro-Archaeology and the Mystical Sciences of Antiquity*. Rochester, VT: Inner Traditions, International, 1989.

Moore, Patrick, and Wil Tirion. *The Cambridge Guide to Stars and Planets*. New York: Cambridge University Press, 2001.

Moura, Ann. *Grimoire for the Green Witch: A Complete Book of Shadows*. St. Paul, MN: Lllewellyn Worldwide, 2004.

Murad, Edmond, and Iwan P. Williams, eds. *Meteors in the Earth's Atmosphere*. New York: Cambridge University Press, 2002.

Murphy, Edward M. *Our Night Sky*. Chantilly, VA: The Great Courses, 2010.

Olcott, William Tyler. *Star Lore: Myths, Legends, and Facts*. Mineola, NY: Dover Publications, 2004.

O'Meara, Stephen James. *Deep-Sky Companions: Southern Gems*. New York: Cambridge University Press, 2013.

Ovid. *The Metamorphoses*. Translated by Charles Martin. New York: W. W. Norton & Company, 2004.

Pickover, Clifford A. *Dreaming the Future: The Fantastic Story of Prediction*. Amherst, NY: Prometheus Books, 2001.

Remler, Pat. *Egyptian Mythology, A to Z*. New York: Chelsea House, 2010.

Ridpath, Ian. *Star Tales*. Cambridge, UK: Lutterworth Press, 1988.

Ridpath, Ian, and Wil Tirion. *The Monthly Sky Guide*. New York: Cambridge University Press, 2006.

Riske, Kris Brandt. *Llewellyn's Complete Book of Astrology*. Woodbury, MN: Llewellyn Publications, 2007.

Robson, Vivian. *The Fixed Stars & Constellations in Astrology*. Abingdon, MD: The Astrology Center of America, 2005.

Rosser, William Henry. *The Stars*. Charleston, SC: Nabu Press, 2012.

Ruggles, Clive L. N. *Ancient Astronomy: An Encyclopedia of Cosmology and Myth*. Santa Barbara, CA: ABC-CLIO, Inc., 2005.

Sasaki, Chris, and Joe Boddy. *The Constellations: Stars & Stories*. New York: Sterling Publishing Co., 2003.

Scalzi, John. *Rough Guide to the Universe*. New York: Rough Guides, 2008.

Schaaf, Fred. *40 Nights to Knowing the Sky*. New York: Henry Holt & Company, 1998.

———. *The Brightest Stars: Discovering the Universe through the Sky's Most Brilliant Stars*. Hoboken, NJ: John Wiley & Sons, 2008.

———. *A Year of the Stars: A Month-by-Month Journey of Skywatching*. Amherst, NY: Prometheus Books, 2003.

Schneider, Howard. *National Geographic Backyard Guide to the Night Sky*. Washington, DC: The National Geographic Society, 2009.

Shadick, Stan. *Skywatcher's Companion: Constellations and Their Mythology*. Custer, WA: Heritage House Publishing, 2011.

Sharpes, Donald K. *Sacred Bull, Holy Cow: A Cultural Study of Civilization's Most Important Animal*. New York: Peter Lang Publishing, 2006.

Simpson, Phil. *Guidebook to the Constellations*. New York: Springer, 2012.

Squire, Charles. *Celtic Myth and Legend: Poetry & Romance*. Mineola, NY: Dover Publications, 2003.

Steiger, Brad. *Totems: The Transformative Power of Your Personal Animal Totem*. New York: HarperCollins Publishers, 1997.

Sykes, Egerton. *Everyman's Dictionary of Non-Classical Mythology*. New York: E. P. Dutton & Co., 1965.

Thompson, Robert Bruce, and Barbara Fritchman Thompson. *Illustrated Guide to Astronomical Wonders*. Sebastopol, CA: O'Reilly Media, 2007.

Thorburn, John E., ed. *The Facts on File Companion to Classical Drama*. New York: Facts on File, 2005.

Thurston, Hugh. *Early Astronomy*. New York: Springer, 1996.

Tresidder, Jack. *The Complete Dictionary of Symbols*. San Francisco: Chronicle Books, 2005.

Turner, Patricia, and Charles Russell Coulter. *Dictionary of Ancient Deities*. New York: Oxford University Press, 2000.

Vamplew, Anton. *Simple Stargazing*. New York: HarperCollins, 2006.

Vamplew, Anton, and Will Gater. *The Practical Astronomer*. New York: DK Publishing, 2010.

Walker, Barbara G. *The Women's Dictionary of Symbols and Sacred Objects*. New York: HarperCollins Publishers, 1988.

Westmoreland, Perry L. *Ancient Greek Beliefs*. San Ysidro, CA: Lee and Vance Publishing, 2007.

Woolfolk, Joanna Martine. *Aquarius: Sun Sign Series*. Lanham, MD: Taylor Trade Publishing, 2011.

——— . *Pisces: Sun Sign Series*. Lanham, MD: Taylor Trade Publishing, 2011.

Yin, Amorah Quan. *The Pleiadian Tantric Workbook*. Rochester, VT: Inner Traditions, 1997.

ONLINE RESOURCES

American Meteor Society: http://www.amsmeteors.org

AppAdvice: http://www.appadvice.com/appguides/show/astronomy-apps

BBC (British Broadcasting Corporation). Transcript of interview with Miranda Green, *Secrets of the Star Disc*, BBC Two, Thursday 29 January 2004, 9 p.m. http://www.bbc.co.uk/science/horizon/2004/stardisctrans.shtml

Google: http://www.google.com/mobile/skymap/

iTouchMap: http://www.itouchmap.com/latlong.html

iTunes: http://www.apple.com/itunes/

NASA (National Aeronautics and Space Administration): http://www.spaceplace.nasa.gov/starfinder3/en/

NPR (National Public Radio). Transcript of April 21, 2013 NPR's Weekend Edition Sunday with host Rachel Martin and guest Kelly Beatty, senior contributing editor for *Sky and Telescope* magazine. http://www.npr.org/blogs/thetwo-way/2013/04/21/178202922/sunday-night-forecast-cloudy-with-a-chance-of-meteors

Sky and Telescope magazine: http://www.skyandtelescope.com

Sky Maps: http://www.skymaps.com

ENDNOTES

1. Leeming, *Goddess*, 78.

2. Moura, *Grimoire for the Green Witch*, 119.

3. Olcott, *Star Lore,* 9–10.

4. Thurston, *Early Astronomy,* 64.

5. Ellis, *Early Irish Astrology. Réalta* (vol 3. n. 3, 1996).

6. Remler, *Egyptian Mythology, A to Z,* 22.

7. National Public Radio, Weekend Edition Sunday with host Rachel Martin, April 21, 2013.

8. Cirlot, *A Dictionary of Symbols,* 242.

9. Artress, *Walking a Sacred Path,* 67.

10. Greer, *The New Encyclopedia of the Occult,* 43.

11. Ridpath, *Star Tales,* 34.

12. Olcott, *Star Lore,* 74.

13. Gimbutas, *Language of the Goddess,* 233.

14. Brady, *Brady's Book of Fixed Stars,* 120.

15. Green, *The Celtic World,* 273.

16. Gallant, *The Constellations*, 24.

17. Olcott, *Star Lore*, 381.

18. Evans, *The History and Practice of Ancient Astronomy*, 102.

19. Ridpath, *Star Tales*, 8.

20. Franklin, *Celtic Animal Oracle*, 34.

21. Anthon, *Latin-English and English-Latin Dictionary*, 81.

22. Ridpath, *Star Tales*, 85.

23. Simpson, *Guidebook to the Constellations*, 42.

24. Brady, *Fixed Stars*, 103.

25. Ibid., 108.

26. Scalzi, *Rough Guide to the Universe*, 284.

27. McIntosh, *Handbook to Life in Prehistoric Europe*, 248.

28. Olcott, *Star Lore*, 294.

29. Brady, *Fixed Stars*, 291.

30. Condos, *Star Myths of the Greeks and Romans*, 106.

31. Conway, *Magickal, Mystical Creatures*, 22.

32. Yin, *The Pleiadian Tantric Workbook*, 26.

33. Remler, *Egyptian Mythology A to Z*, 22.

34. Agrippa, *Three Books of Occult Philosophy*, 396.

INDEX

Stars by Traditional Names

General Index

To Write to the Author

If you wish to contact the author or would like more information about this book, please write to the author in care of Llewellyn Worldwide Ltd. and we will forward your request. Both the author and publisher appreciate hearing from you and learning of your enjoyment of this book and how it has helped you. Llewellyn Worldwide Ltd. cannot guarantee that every letter written to the author can be answered, but all will be forwarded. Please write to:

Sandra Kynes
℅ Llewellyn Worldwide
2143 Wooddale Drive
Woodbury, MN 55125-2989

Please enclose a self-addressed stamped envelope for reply,
or $1.00 to cover costs. If outside the U.S.A., enclose
an international postal reply coupon.

Many of Llewellyn's authors have websites with additional information and resources. For more information, please visit our website at http://www.llewellyn.com.

Sea Magic
Connecting with the Ocean's Energy
SANDRA KYNES

Purification, regeneration, transformation…the ocean has long been known for its extraordinary powers. Sandra Kynes invites you to explore the sea's mystical energies. This spiritual voyage can help you find balance in your life or guide you toward a "sea change" of your own.

Sea Magic is for everyone. You don't have to live on the coast or follow any specific spiritual path to tap into the unique energy of the ocean. Kynes offers meditations and exercises to help you center yourself, explore emotions, and find your place in the vast web of life. Dive into inner worlds of imagination and creativity. Choose a sea fetch (totem animal) to take you on a shamanic journey. Build an altar to focus your intentions and learn the rhythms of the moon.

From working with sea deities to divining with seashells, *Sea Magic* offers ample ways to enhance your life and open up to divine guidance.

978-0-7387-1353-3, 240 pp., 6 x 9 $15.95